Treatment of Cooling Water

Aquaprox (Ed.)

Treatment of Cooling Water

Springer

Aquaprox
6 rue Barbés
92305 Levallois
France
E-mail: fpinatel@protex-international.com
aquaprox@protex-international.com

ISBN 978-3-642-01984-5 e-ISBN 978-3-642-01985-2

DOI 10.1007/978-3-642-01985-2

Library of Congress Control Number: 2009934314

Originally published as Aquaprox, Traitement des eaux de refroidissement, Lavoissier 6.2006

Typesetting: Data supplied by the authors

Production: Scientific Publishing Services Pvt. Ltd., Chennai, India

Cover Design: WMX Design, Heidelberg, Germany

Printed in acid-free paper

9 8 7 6 5 4 3 2 1

springer.com

Foreword

The **AQUAPROX** Company is the "water treatment" subsidiary of the **PROTEX INTERNATIONAL** Group.

Specialized for more than 70 years in fine chemistry, the **PROTEX INTERNATIONAL** Company exercises its activities in the following fields : chemistry, textiles, electronics, agro-food, pharmacy, perfumes, and, with AQUAPROX, water treatment.

AQUAPROX is defined to be a service company for industrials in the water treatment field.

AQUAPROX provides its technical competence and raw materials synthesis technology through to the guarantee of results associated with outsourcing contracts signed with world leaders in their markets.

The range of products marketed by **AQUAPROX** include the following families:

- For cooling circuits: antiscaling, dispersing, anti-corrosion, bactericide, algicide, biodispersing agents;
- For boilers: oxygen scavenger, alkalizing agent, phosphatizing agent, antiscaling agent;
- For wastewater treatment plants: coagulant, flocculant, anti-foam agent, biological products generating reductions in the volume of sludge, eliminating unpleasant odours, treating of hard COD.

The publication titled *Cooling Water Treatment* summarizes the 20 years of experience acquired in the field.

This publication, written by the successive water treatment professionals in **AQUAPROX**'s commercial management (J.-C. Berce, D. Billon-Lanfray), describes all the problems encountered in the management of industrial cooling circuits.

It offers concrete solutions.

It is practical but not superficial, proposing individual reflections rather than exhaustive approaches.

It is intended for circuit operators as well as buyers concerned about obtaining a critical view on technical responses to requests for proposals for both environmental managers and directors of units or industrial sites.

It concerns all industrial sectors requiring cooling; the main sectors are:

- Energy production;
- Steel industry;
- Paper industry;
- Agro-food industry;
- Electronics.

It focuses on simple study, follow-up and control elements which will allow:

- Generating gains: water and energy savings;
- Extending the life of the industrial equipment: corrosion control;
- Making production reliable and regulating it: control of scaling, clogging and bacterial development;
- Proposing products in terms of active materials the least toxic as possible for man and the environment;
- Guaranteeing results in relation with safety, health and environmental standards.

Cooling products while they are being made or at the end of their manufacture are a part of the process associated with most industrial activities.

For instance, semi-open cooling circuits are conventionally present in the following industrial sectors:

- Energy production; to ensure vapor condensation to cool motors, pumps, etc;
- Steel industry, in all the steel manufacturing phases:
- . blast furnaces, to cool the blast furnace itself, hot blast valves, the casting top, all elements subjected to high temperature may be damaged and even their mechanical resistance may break down;
- . steel mill: oxygen pipes, ingot molds, etc.;
- . hot rolling mills : furnace beams, rollers, rotating parts;
- . coke plant: ovens, distillation columns, etc.;
- Agro-food industry: breweries, sugar mills, distilleries, dairies, starch factories, canneries, slaughter houses, fruit juices, ice cream production. In all these industries, there are cooling phases which are indispensable to condition the finished product before shipping or to regulate the temperature in pasteurization and sterilization operations;
- Paper industry: cooling of black liquors, motors, compressors;
- Refining: to cool cracking units, distillation columns, etc.;
- Chemistry and pharmaceutical industry: to cool reactors where organic or inorganic syntheses take place;
- Plastics transformation: to cool dies;
- Automotive industry: production of motors (aluminum foundries), manufacture of gearboxes (machining), etc.;
- Electronic parts production: air conditioning of clean rooms;
- Wastewater treatment plants: cooling of compressors.

Cooling circuits are also installed in the air conditioning systems of most office buildings, shopping centers, hospitals, airports, hotels.

In each of these circuits, potential problems of scale depositing, corrosion, legionella development and so on exist.

AQUAPROX
6, rue Barbès
92300 Levallois, France
Phone : +33 (0)1 41 34 14 00
Fax : +33 (0)1 41 34 14 34
E-mail : postmaster@protex-international.com
Internet : www.aquaprox.com

Preface

When, over twenty-five years ago, I began my career as a young engineering graduate, I discovered many things about the use of cooling water.

A thermodynamicist by training, at the time I considered water to be a thermal fluid whose primary function was to evacuate heat energy and optimise the efficiency of installations. I very quickly became disillusioned when I saw the efficiency of water cooling towers and heat exchangers decrease spectacularly after the formation of a deposit a few millimetres thick of carbonated scale.

This early experience struck me to such an extent that, several years after leaving university, I went back to the lecture hall to get a master's degree in industrial chemistry, followed by a doctorate in science processes. Later, industrial uses of water became an important part of my professional life and have always remained a real passion for me.

Now, I truly hope to be able to share this with my students.

This book will add to a relatively limited series of works which deal with the problematics of industrial water, by including all the aspects by which the practitioner is confronted: scaling, fouling, corrosion and biological developments. It also serves as a reminder of the basic concepts of water treatment, from physico-biochemical characterisation to the succinct description of the various purification processes. Finally, it concentrates on the various cooling water treatment processes, which is the main objective of the book.

When I started out in the world of industry, I would have liked to have had a work like this available to me. It's a book for general use, which, over a few evenings of attentive reading, would have given me an overview of the problematics linked to the use of cooling water described within, which are so important.

Léon DUVIVIER – PhD
Professor – Catholic University of Louvain-la-Neuve
Technical Director Loborelec, Group Suez

Contents

FIRST PART

SECOND PART

FIRST PART

A – Natural Water

Introduction

Water exists in nature in enormous quantities, in the liquid, gaseous or solid state. It is estimated that it covers 70 % of the earth's surface.

The greatest part of water is constituted by salt water of the seas and oceans and, finally, fresh water represents only a relatively small proportion of the total quantity of water.

Throughout its cycle (evaporation from oceans and seas, atmospheric precipitation, runoff into rivers and a return to the oceans), the composition of water varies continually and it may be more or less distanced from "pure substance" water, known by physicists and chemists.

1

Pure Water

It seems to be useful to remind the physical and chemical characteristics of water, and indicate its variation depending on certain parameters.

1.1 Chemical Formula

As we see it from the formula H_2O and the molecular weight 18, water is constituted by two atoms of hydrogen and one atom of oxygen.

The presence of hydrogen isotopes, particularly of the deuterium of mass 2 (symbol D) in natural hydrogen in the molecular proportion of 1/6000 makes that water contains a non-negligible proportion of heavy water D_2O and semiheavy water HDO, while other isotopes present in it are negligible.

1.2 Physical Properties of Water

1.2.1 Density

Liquid water presents a maximum specific weight at 4 °C: 0.99997, which corresponds to a density of 1.00000.

Temperature °C	Density of water
0°	0.999
4°	1.000
20°	0.998
50°	0.998
110°	0.958
150°	0.917
200°	0.863
250°	0.794
300°	0.70

Variation in pH of pure water depending on the temperature:

0 °C	pH 7.5
50 °C	pH 6.6
100 °C	pH 6.1
160 °C	pH 5.8
200 °C	pH 5.7
250 °C	pH 5.5

This decrease in pH depending on temperature is an extremely important point that we tend to forget.

In other words, pure water overheated at 180 °C, as, for example, water for heating of housing complexes, has a pH around 5.75. It is therefore acid, thus aggressive.

In order to obtain, at 180 °C, a neutral pH or slightly alkaline, it is necessary that the pH at low temperature is clearly alkaline and close to 10.

Moreover, these values are modified when water contains dissolved substances likely to bring the ions H^+ or OH^-.

For further information:

- natural water has a pH value of 5 to 9;
- fruit juices have an acid pH of 2.2 (lemon);
- decinormal sulfuric acid has a pH value of 1.2.

1.3 Thermodynamic Properties

- Melting point: 0 °C, not very sensitive to the pressure variation.
  Heat of fusion: 330 kJ·kg $^{-1}$ (or 80 kcal/kg).

Boiling point: 100 °C under the pressure of 760 mm of mercury. It varies depending on the pressure.

Heat of vaporization: 2250 kJ·kg $^{-1}$ (539 kcal/kg) a 20 °C. It varies also depending on the temperature. (See the table n° 1).

1.4 Water-Air System

Whenever we work with the water-air mixture (washers, cooling towers), the input air and output air are not in the same hygrometric state and it is necessary to know the quantity of water contained in the air depending on the temperature and the state of saturation (See tables n° 2, 3 and 4, p. 49 and 50).

2

Natural Water

2.1 Various Natural Waters

The liquid water which is formed by the condensation at high altitude is relatively pure and, before reaching the ground, it dissolves gases and absorbs soluble and insoluble compounds present in the air. This water is generally saturated with oxygen, contains the carbon dioxide and can be contaminated by gases issued from industry: nitrogen oxide, hydrogen sulphide, sulphurous gas. It can also contain organic matters and minerals coming from the ground in the form of dust.

When it reaches the ground, many components of the Earth's crust can be dissolved in rain water during the trickling or infiltration into the permeable geological layers, and this is connnected with the content of the gas dissolved, which helps the dissolving of the minerals, in particular, the carbon dioxide.

Natural fresh waters can be classified into three categories:

- rain water;
- surface water: flowing water, water pools, lakes and reservoirs of surface;
- groundwater: coming from springs, deep and shallow wells and water tables.

Rain water is hardly ever used, however, there are current experiments in order to collect it and use it as make-up water in the cooling circuits. Its low mineralization constitutes a distinctive feature that implies a specific use at each industrial site.

Surface waters are in general characterized by a high concentration of oxygen and a low carbon dioxide content, since they are in equilibrium with the air which contains 21 % of oxygen and only about 0.03 % of carbon dioxide.

The surface waters are often turbid because of the presence of clay and organic wastes issued from vegetation, agriculture, and they are, in general, rich in microorganisms. They can be polluted by the wastes produced from human activities: city sewers, products issued from agriculture and breeding, industrial wastes...

Water tables/aquifers are a composition highly variable depending on surrounding soils; they have, in general, lost their turbidity and have acquired a greater or lesser hardness. Often, there are contaminations coming from the surface. The same goes for spring water whose composition is the same as that of the water tables, from which it is issued.

Deep waters are generally much more mineralized and often much richer in carbon dioxide dissolved, of which the solubility increases depending on the pressure to which it is exposed (at which it is situated).

We note as reminder that the sea water is not covered by the scope of the current book.

Its use as the make-up water in once-through cooling circuits has been developping at the seaside. Let us just to remind that it comes from the geological concentration of all these soluble substances, and it is mainly characterized by its high concentration in alkaline halogenides and alkaline-earth elements.

2.2 Impurities of Water

By "impurities of water" we mean all that is not pure substance H_2O.

2.2.1 Suspended Solids

It is a matter of solid particles of the density superior to that of water and remaining suspended only because the water is in movement: the higher the speed of the water, the larger can/might be the suspended particles.

Their nature can be either mineral: it is the case of sand, silt and clay; or biological: it is the case of the bacteria and organic matters.

The suspended particles settle or sediment as soon as the water is no more agitated or less agitated; for water at rest, the settling velocity is calculated according to the Stokes' law, having the following formula:

$$V = \frac{1/18\ g\ D^2\ (d_1 - d_2)}{\eta}$$

where V: terminal velocity or settling velocity of the particle in cm/s

D: its diameter in cm

d_1: density of the particle in g/cm^3

d_2: density of the liquid in g/cm^3

η: dynamic viscosity of the liquid (in pascal seconds)

The Stokes' law calculates the limit/threshold settling velocity of the particles assimilated to spheres in a fluid having a number of Reynolds (N_{Re}) inferior to 0,3.

For particles of the density of 2.65, we get the following table:

Diameter of the suspended particella Particle	Nature of the Particle	Time required for running 1 m
10 mm	Gravel	1 second
1 mm	coarse sand	10 seconds
0.1 mm	fine sand	2 minutes
0.01 mm	Silt	2 hours
1 µm	Bacteria (different density)	4 days
0.1 µm	Clay	2 years
0.01 µm	Colloid	10 years

2.2.2 Dissolved Solids

Dissolved solids which are most commonly found are the following:

- biocarbonates:

— calcium bicarbonate $Ca(HCO_3)_2$
— magnesium bicarbonate $Mg(HCO_3)_2$
— sodium bicarbonate $NaHCO_3$
— ferrous bicarbonate $Fe(HCO_3)_2$
— manganese bicarbonate $Mn(HCO_3)_2$

- chlorides:

— calcium chloride $CaCl_2$
— magnesium chloride $MgCl_2$
— sodium chloride $NaCl$

- sulphates:

— calcium sulphate $CaSO_4$
— magnesium sulphate $MgSO_4$
— sodium sulphate Na_2SO_4

- silicates:

— sodium cilicate Na_2SiO_3
— silicic acid H_2SiO_3

The following ones are found as well:

— fluorides F^-
— nitrates NO_3^-
— nitrites NO_2^-

2.2.3 Dissolved Gases

The main gases found dissolved in natural water are the following:

— oxygen O_2
— nitrogen N_2
— carbon dioxide CO_2
— hydrogen sulphide H_2S.

The quantity of gas dissolved in a liquid is proportional to:

— the (proper) solubility of the gas in water
— the concentration of the gas in atmosphere in contact with water.

It is the case of oxygen and nitrogen.

As for the acid anhydrides such as carbon dioxide and hydrogen sulphide H_2S, to the previously defined solubility are added the hydration and ionization, which increase their solubility.

According to the reactions:

$$CO_2 + H_2O \leftrightarrow HCO_3 + H^+ \leftrightarrow CO_3^- + 2H^+$$

the balance is at stake depending on the pH. It is the same for H_2S.

It results in the following table of solubility:

Solubilities of gases in water (litres of gas per litre of water) in an atmosphere of pure gas (under a pressure of 1 bar, at 0 °C)

	Temperature °C							
	0	5	10	15	20	25	30	35
Oxygen	0.0489	0.0429	0.0380	0.0342	0.0310	0.0283	0.0261	0.0244
Atmospheric Nitrogen	0.0235	0.0209	0.0186	0.0169	0.0155	0.0143	0.134	0.0126
Carbon Dioxide	1.713	1.424	1.194	1.019	1.878	0.759	0.665	0.592
Hydrogen Sulphide	4.670	3.977	3.399	2.945	2.582	2.282	2.037	1.831

It results from this table that the « air » dissolved in water is richer in oxygen than the atmospheric air is. Moreover, the solubility of the gases decreases with the increase in temperature and it increases depending on the pressure increase.

Solubility of oxygen in water equilibrated with water-saturated air at atmospheric pressure.

Temperature (°C)	Oxygen (mg/l)
0	14.8
5	12.8
10	11.3
15	10.2
20	9.2
25	8.4
30	7.6
35	7.1
40	6.6
45	6.1
50	5.6

In practice, the concentration of dissolved oxygen is particularly significant in surface water, especially in rivers where the dissolving is activated by agitation, and this is all the more so since the water is less rich in organic matters subject to biodegradation.

The carbon dioxide, whose extent is not significant in surface water, becomes very significant in underground water, also in relation with the pressure which influences it there; finally, an effervescence by decompression can occur there (mineral water).

3

Analysis of Natural Water

3.1 Units Chosen

We have seen in the list of dissolved salts, besides halogens found also among dry solids in water, the bicarbonates which exist only in solution: these are, for instance, calcium bicarbonate, magnesium bicarbonate and ferrous bicarbonate. These salts are not insoluble, and they are found in dry matter in the form of carbonate, as for magnesium and calcium, according to the reaction:

$$Ca\,(HCO_3) \rightarrow CaCO_3 + CO_2 + H_2O$$

$$Mg\,(HCO_3)_2 \rightarrow MgCO_3 + CO_2 + H_2O$$

The ferrous carbonate is found in the form of ferric hydroxide by carbonate removal followed by oxydation.

$$Fe\,(HCO_3)_2 \rightarrow Fe\,(OH)_2 + 2CO_2$$

$$2\,Fe\,(OH)_2 + 1/2\,O_2 + H_2O \rightarrow 2\,Fe(OH)_3$$

In the case of the sodium bicarbonates, the carbonate or bicarbonate is obtained depending on the temperature of evaporation.

Moreover, we must not forget that, as these salts are present in dilute medium, they are highly ionized, as well as the fact that we have the presence of ions in water:

$$Cl^-, F^-, HCO_3^-, CO_3^{--}, Ca^{++}, Fe^{++}, Mg^{++}, Na^+$$

and, in the practice of the analysis, it is these ions that we measure out; the analysis will thus be expressed in ions, either in mg/litre, which accounts for the mass dissolved, or in milliequivalents per litre, which account for the number of ions present. This allows interpreting rationally the analysis, as the sum (in number) of the cations is equal to the sum of the anions, since the solution is electrically neutral.

There are four common expressions of results:

a) concentration in an ion in mg per litre,

b) concentration in an ion in meq/litre, these two being connected by the following relation:

$$\text{meq/l} = \frac{\text{mg/l} \times V}{PM},$$

where PM: molecular weight of the ion,
V: its valence.

It must be remarked that the meq/litre are equal to the equivalents per million (epm) of Anglo-saxons and the mg/l can be assimilated to ppm (weight).

Ion	Molecular weight	Valence	Weight in mg of milliequivalent
Ca^{++}	40.10	2	20.05
Mg^{++}	24.30	2	12.15
Na^{+}	23.00	1	23.00
OH^{-}	17.00	1	17.00
CO_3	60.00	2	30.00
$HCOs^{-}$	61.00	1	61.00
Cl	35.46	1	35.46
SO_4^{--}	96.06	2	48.03

c) concentrations expressed partly per million of $CaCO_3$, sometimes used so as to express all the ions in $CaCO_3$, because the calcium carbonate is almost always the most significant deposit.

d) concentrations expressed in degrees:

The degree is an arbitrary unit which corresponds, in France, to 0.2 meq/litre. Other countries also have their definition of the degree which does not correspond to the French degree (see the table n° 5, pages 57 and 58).

3.2 Elements Measured and Interpretation of Results

The typical analysis of water has in general the following feature:

Total hardness (in the form of $CaCO_3$) 115 ppm
Calcium hardness (in the form of $CaCO_3$) 70 ppm

Magnesium hardness (in the form of $CaCO_3$)	45 ppm
Total alkalinity (in the form of $CaCO_3$)	150 ppm
Chlorides (in the form of Cl^-)	30 ppm
Sulphates (in the form of SO_4^{--})	15 ppm
Silica (under the form of SiO_2)	5 ppm
pH	7.2.

We will examine the characteristics that each element of the analysis represents.

3.2.1 Hardness of Water: Titration for Hardness (TH)

Water, when it contains calcium and magnesium salts, is called "hard"; it results in undesirable properties: possibility to form hard deposits through heating of water, difficulties in washing opérations using precipitation soaps, insoluble metallic soaps, etc.

Fife diffferent titres for hardness can be defined:

— *TH or total titration for hardness (or total hardness),* corresponds to the totality of ions of calcium and magnesium present in water.

— *Calcium TH (or calcium hardness),* corresponds to the global content of calcium salts in water.

— *Magnesium TH (or magnesium hardness),* correspond to the global content of magnesium salts in water.

We have the following relation:

$$TH = \text{calcium TH} + \text{magnesium TH}.$$

When boiling water, we see that it lets deposit calcium and magnesium carbonates, and the water, after this treatment, loses a part of its hardness; for this reason we define:

— *Permanent TH (also called non-carbonate hardness):* it is one that remains after boiling of water in standardized conditions; it is called permanent hardness or non-carbonate hardness; it corresponds to chlorides and sulphates of calcium and magnesium.

— *Temporary TH:* it is one that disappears after the boiling; it corresponds to carbonates and bicarbonates of calcium and magnesium; it is called temporary hardness or carbonate hardness.

We have the following relation:

$$TH = \text{temporary TH} + \text{permanent TH}.$$

Expression of the TH: they can be expressed in meq/litre, in mg of $CaCO_3$ per litre, or in degrees (see the table n° 5).

3.2.2 Alkalinity: Titres

If we titrate high pH water by a strong acid (H_2SO_4 or HCl, for instance), we would obtain a neutralization curve presenting two points of neutralization, which correspond: the one to the methyl orange colour change and the other – to that of phenolphthaleine. To each of these colour changes corresponds a specific alkalinity.

► *PA or phenolphthalein alkalinity (p-alk), corresponding to French TA ("titre alcalimétrique")*

It is the first point of neutralization at pH 8.3 characterized by the passage of the phenolphthaleine from red into colourless, and it corresponds to the neutralization of free bases, sodium hydroxide, lime:

$$Ca(OH)_2 + 2\ HCl \rightarrow CaCl_2 + H_2O$$

and to the half-neutralization of the carbonates (passage to bicarbonate):

$$Na_2CO_3 + HCl \rightarrow NaHCO_3 + NaCl$$

► *MA or Methylorange alkalinity (m-alk), corresponding to French TAC ("complete TA")*

It is the second point of neutralization at pH 4.5 characterized by the passage of the methylorange from yellow to pinkish; it corresponds to the neutralization of the bicarbonates, that is to say, half of the carbonates and bicarbonates which were initially present.

3.2.3 Interpretation of Alkalinities

Knowledge of the PA (p-alk) and MA (m-alk) allows calculating concentrations in:

— hydroxides	OH^-
— bicarbonates	$HCO3^-$
— carbonates	$CO3^{--}$

If the PA and MA are equal, the whole alcalinity is due to the hydroxides.

If the PA is inferior to the MA, it means the presence of carbonates and hydroxides. This means that the PA is equal to all the hydroxides, plus half of the carbonates, and the MA is equal to all the hydroxides and all the carbonates. Therefore:

$PA = (OH^-) + 1/2\ (CO_3^{--})$
$MA = (OH^-) + (CO_3^{--})$
$(OH^-) = 2\ PA — MA$
$(CO_3^-) = 2\ (MA — PA)$

If the PA is equal to half of the MA, the whole alcalinity is due to the carbonates.

If the PA is equal to less than half of the MA, it means the presence of carbonates and bicarbonates. This means that the PA is equal to half of the alkalinity due to the carbonates, and the MA is equal to the alkalinity due to the bicarbonates, plus the alkalinity due to the carbonates. Therefore:

$(CO_3^{--}) = 2\ PA$
$(HCO_3^{--}) = MA - 2\ PA$

If the PA is nil, the MA is only due to the bicarbonates. To summarize:

If	OH^-	CO_3^{--}	HCO_3^-
PA = MA	MA	0	0
PA > 1/2 MA	2PA—MA	2 (PA—MA)	0
PA = 1/2 MA	0	MA	0
PA < 1/2 MA	0	2 PA	MA—2PA
PA = 0	0	0	MA

► *Acid water cases*

In the case where the pH of water is < 4.5, we can define the strong acid content (SAC).

In this case: PA = MA = 0

The SAC is due only to dissolved carbon dioxide; there are strong acids of non-residual origin (in general, added or due to a passage on an ion exchange resin). There is no carbonate, no bicarbonate and no free bases here, but only the salts dissolved.

► *Relation between hardness (TH) and alkalinity (PA and MA)*

If MA = 0: The whole hardness is non-carbonate (permanent).
If MA < TH: TH — MA = non-carbonate hardness (permanent)
MA = carbonate hardness (temporary).

If MA = TH: The whole hardness is carbonate.
If MA > TH: The whole hardness is carbonate
MA — TH = alcalinity due to the monovalent salts, sodium and potassium.

These relations tell us something on possible treatements of water. The carbonate hardness can be eliminated by a treatment using pure lime. The non-carbonate hardness can also be eliminated, but requires the use of lime and anhydrous sodium carbonate (this technique is not used in practice).

▶ *Significance of pH*

According to the conventional definition, « the pH is the decimal logarithm of the inverse (reciprocal) of hydrogen ion concentration ».

$$pH = \log 1/ [H^+]$$

The pH is the measure of the acidity or alcalinity of a solution, by means of a scale varying from 0 to 14. The pure water, considered to be neutral, has a pH of 7. The values inferior to 7 mean that the solution becomes more and more acid when pH decreases until 0, and the values superior to 7 mean that it becomes more and more alkaline when it increases until 14.

In order to illustrate these values, let us give some examples: orange juice and fruit drinks have a pH of about 3.5; lemon juice has a pH of 2.2, and decinormal sulphuric acid has a pH of 1.2. As for the alkaline products, sodium bicarbonate has a pH of about 8.4; magnesia milk has a pH of 10.5, ammonia for household use – 11.5 and decinormal sodium hydroxide has a pH of 13.0.

An important thing to bear in mind is that the pH scale is logarithmic. A pH of 5 is 10 times more acidic than a pH of 6; a pH of 4 is 10 times more acidic than a pH of 5, or 100 times more acidic than a pH of 6.

▶ *Strong acid salts (SAS) in water*

Up to the present, we have classified by acidimetry only salts of weak acids and bases, that is to say, the ions OH^-, HCO_3^- and CO_3^{--}, because they are alkaline in presence of phenolphthaleine or methylorange.

In natural water, there also are salts of strong acids, sulphates, chlorides, nitrates, etc.

We can either measure separately and specifically different ions, which we do generally at least for chlorides and sulphates, or measure salts of strong acids: the determination is made in two times.

Firstly, we pass water to be analyzed on a cation exchange resin of the cycle H, which releases all the acids, and secondly, we measure the acidity appeared at the methylorange colour change.

The result can be expressed in meq/l or in French degree:

1 French degree = 0.2 meq/l.

We also measure silica which is present in certain water. It is expressed in SiO_2, and never in the ionic or neutral form and is not included in the line of account in the calculations, because we de not know whether it is found in the form of ion or colloid; indeed, even in the ionic form, the silicic ion content is too small/low to be measured by titrimetry; it represents only a minute part in the weight of dissolved salts.

3.2.4 Global Interpretation of the Analysis: Ionic Balance and Weight Balance

3.2.4.1 Ionic Balance

The basic principle of the ionic balance is that, water being practically neutral, the sum of the cations is equal to the sum of the anions, while pH does not intervene, as its scale is logarithmic.

For example, highly acidic water having a pH of 4 has a concentration of ion g/litre = $10^{-4}H^{+}$, or:

1/10,000 mg/1, which is negligible in terms of weight.

We can refer again to the typical analysis of the page 14 and express it in meq per litre.

	mg/l	meq/l	Degrees
TH, total hardness in $CaCO_3$	115	2.30	11.5
Calcium hardness in $CaCO_3$	70	1.40	7
Magnesium hardness in $CaCO_3$	45	0.90	4.5
MA, total alkalinity in $CaCO_3$	150	3.00	15
Chlorides in Cl^-	30	0.845	0.42
Sulphates in SO_4^{--}	15	0.313	0.16
Silica in SiO_2	5	—	
pH	7.2	—	

First, we remark that:

As the MA is superior to the TH, the whole hardness is carbonate and the difference MA — TH = 3.0 — 2.3 = 0.7 indicates the presence of monovalent carbonate ions (which do not intervene in hardness).

Moreover, as the pH is 7.2, the PA is definitely nil, we deduce from it that the alkalinity is entirely due to the bicarbonates.

We can deduce from it the alkaline ions' content:

a) alkaline bicarbonates 0.7 meq/1
b) alkaline chlorides 0.845 meq/1
c) alkaline sulphates 0.313 meq/1

Thus, we have in total 1.858 meq/g of alkaline ions that we would suppose to be the ions of Na, after the supplementary analysis of K, Na, NH_4, etc.

The ionic balance is thus established in meq/1:

Anions		Cations	
Hydroxides	0	Calcium	1.40
Carbonates Bicarbonates	0	Magnesium	0.90
Chlorides	3.00	Alkaline metals	1.858
	0.845		
Sulphates	0.313		
	4.158 meq/1		4.158

3.2.4.2 Weight Balance

	meq/1	Weight of meq	Concentration of free mg
Bicarbonates	3.0	61	183
Chlorides	0.848	35.5	30
Sulphates	0.313	48	15
Calcium	1.40	20	28
Magnesium	0.90	12.1	11
Alkaline (sodium)	1.858	23	43
Silica			5
Dissolved salts			315

3.2.4.3 Free Carbon Dioxide Content

As the carbonic acid is a weak acid, it is not completely ionized even at low concentration, and a part of CO_2 is present in the form of dissolved CO_2, while this quantity is all the more significant since the pH is low and the MA is high.

The graph of the following page indicates for the relevant example: pH of 7.2 and MA of 150 expressed in ppm of $CaCO_3$, a concentration of free CO_2 of 20 ppm in equilibrium with the solution.

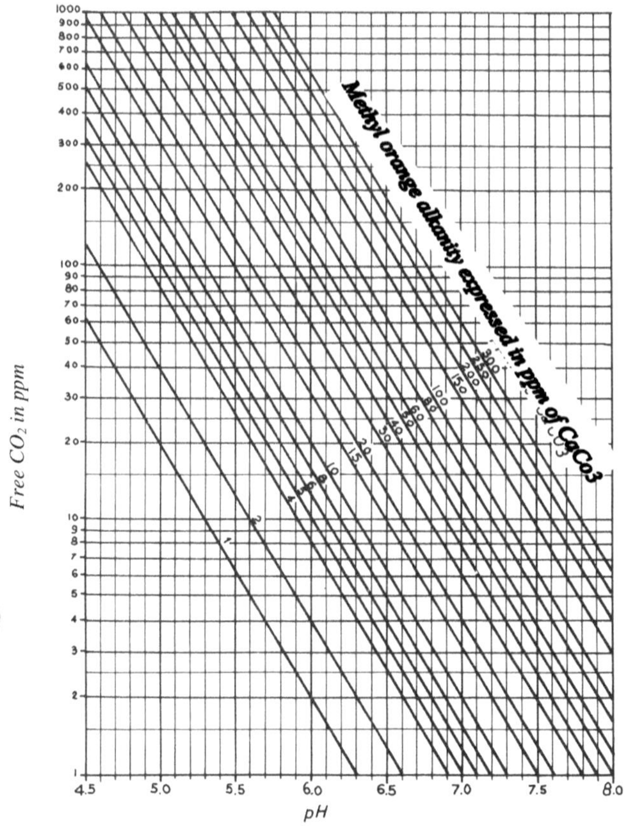
Free CO_2 in ppm
1000
900
800
700
600
500
400
300
200
100
90
80
70
60
50
40
30
20
10
9
8
7
6
5
4
3
2
1
4.5
5.0
5.5
6.0
6.5
7.0
7.5
8.0
pH
Methyl orange alkanity expressed in ppm of CaCo3
1
2
4
6
8
10
15
20
30
40
50
60
80
100
150
200
250
300

15

3.3 Methods Chosen and Equipment Required

We will describe in general two different methods for the same analysis.

3.3.1 Alkalinity

3.3.1.1 Principle

The alkalinity is determined by titration with a strong acid. It is due to a large number of various ions among which the most common are the following:

— free bases	(ions OH^-)
— bicarbonates	(ions HCO_3^-)
— carbonates	(ions CO_3^{--})

Other ions, phosphates and silicates can also contribute to the alkalinity.
Two determinations have been made successively on the same test:

— the alkalinity (PA) or p-alkalinity is determined by the phenolphthaleine colour change at the pH of 8.3;
— the total or complete alkalinity (MA) or m-alkalinity is determined by the methylorange colour change at the pH of 4.3.

3.3.1.2 Reagents and Equipment Needed

► *Equipment*

— burette of 25 ml
— pipette of 50 ml
— beaker or Erlenmeyer flask of 250 ml
— agitator made of glass or magnetic stirrer

► *Reagents*

- sulphuric acid	N/50
- methylorange	in solution
- phenolphthaleine	in solution

3.3.1.3 Procedure

Pipette in a beaker of 250ml, exactly 50ml of water to be analysized; add 4 drops of phenolphthaleine, and if the phenolphthaleine turns red, then titrate by H_2SO_4 N/50 until the decolourizing; or V_1 the volume of the acid used, add 4 drops of methylorange, titrate by H_2SO_4 N/50 until the colour change from yellow to salmon pink; or V_2 the volume of the acid used for obtaining the colour change (V_2 is definitely superior to V_1, since V_2 comprises V_1).

3.3.1.4 Expression of Results

PA = alkalinity,	N = normality of the acid,
expressed in meq/litre	PA = V_1 x N x 1,000 /ml of sample
expressed in French degree	PA = V_1 x N x 5,000 /ml of sample
expressed in ppm $CaCO_3$	PA = V_1 x N x 50,000 /ml of sample

MA = methylorange (complete) alkalinity,
expressed in meq/litre MA = V_2 x N x 1,000 /ml of sample
expressed in French degree MA = V_2 x N x 5,000 /ml of sample
expressed in ppm $CaCO_3$ MA = V_2 x N x 50,000 /ml of sample

Remarks:

a) Clearness of colour changes

The colour change of the phenolphthaleine is clear, there is no problem.

The colour change of the methylorange is more delicate to assess. We would get easily too strong results, if we bring the colour change to clear pink instead of salmon pink; a colour change standard can be obtained by adding to 100 ml of distilled water a drop of methylorange and a tip of a spatula of potassium acid phthalate.

b) The colour changes are clearer, when we place the beaker on a white base.

c) In the case of too colourful water, it is necessary to carry out the measurement potentiometrically at pH of 4.3 for the MA and 8.3 for the PA.

d) We can also carry out these two measurements, using hydrochloric acid N/50.

3.3.2 Free Mineral Acidity

3.3.2.1 Principle

It is quite rare that natural water is found acid in the presence of methylorange; on the other hand, the case of acidity can be found in water that is already treated or contaminated by strong acids (HCl, H_2SO_4, HNO_3).

The measurement of the free mineral acidity is made by the alkalimetry using sodium carbonate in the presence of methylorange.

3.3.2.2 Reagents and Equipment Needed

► *Equipment*

- burette of 25 ml
- pipette of 50 ml
- beaker or Erlenmeyer flask of 250 ml
- agitator made of glass or magnetic stirrer

► *Reagents*

- sodium carbonate N/50
- methylorange in solution

3.3.2.3 Procedure

Pipette in a beaker of 250 ml, exactly 50 ml of water to be analyzed; add 4 drops of methylorange; titrate by the sodium carbonate N/50 until the colour change from pinkish to yellow, or/given V the volume of sodium carbonate used.

3.3.2.4 Expression of Results

TAF = strong acids content/titre, N = normality of sodium carbonate,
expressed in meq/l SAC = V x N x 1,000 /ml of sample
expressed in French degree SAC = V x N x 5,000 /ml of sample
expressed in ppm $CaCO_3$ SAC = V x N x 50,000 /ml of sample

3.3.3 Total Hardness

3.3.3.1 Principe

Any natural water contain, at higher or lower concentration, salts of alkaline-earth metals, calcium and magnesium, which contribute to its hardness.

The total hardness corresponds to the sum of all the alkaline-earth ions present independently from anion present.

The measurement is based on the determination of quantity of the total calcium and magnesium by the ethylene diamine tetra-acetic acid (EDTA) in the presence of a magnesium ion indicator. The calcium is chelated by EDTA more (strongly) than the magnesium is, however, the colour change is obtained only when all the calcium and all the magnesium are chelated. The other alkaline-earth ions, strontium and barium are also measured.

3.3.3.2 Reagents and Equipment Needed

► *Equipment*
- burette of 25 ml
- pipette of 50 ml
- beaker or Erlenmeyer (flask) of 250 ml
- agitator made of glass or magnetic stirrer
- measuring pipette of 5 ml

► *Raagents*
- buffer pH 10
- Eriochrome Black T
- EDTA 2 Na titrated solution M/100 = 0,01 M

3.3.3.3 Procedure

Pipette in a beaker of 250 ml, exactly 50 ml of water to be analyzed; add 5 ml of buffer pH 10; add a tip of a spatula of Eriochrome Black T; titrate by the EDTA 0,01 M until the colour change from winy red to bright blue.

3.3.3.4 Expression of Results

Total TH = Total hardness, N = molarity of EDTA,
expressed in meq/litre Total TH = V x N x 2,000 /ml of sample
expressed in French degree Total TH = V x N x 10,000 /ml of sample
expressed in ppm $CaCO_3$ Total TH = V x N x 100,000 /ml of sample

The titration must be finished slowly in order that you have a correct measurement. It is also possible to use heating to 40 - 50° C to accelerate the speed of the reaction.

The colour change is clear only if the water to be analyzed contains magnesium, which is the most common case. In the case where the colour change is not clear due to the lack of magnesium in the water, it suffices to add, before titration, about 1 ml of the complexe of magnesium of the EDTA M/100.

3.3.4 Calcium Hardness

3.3.4.1 Principle

The purpose of this method is to measure only the calcium among all the alkaline-earth elements present. As in the case of the total hardness, the measurement is made using complexometric titration with the EDTA, while the indicator chosen is a calcium ion indicator insensitive to magnesium. In these conditions, since the calcium complex is the most stable one, the colour change occurs before magnesium interferes. Similarly, strontium and barium do not interfere.

3.3.4.2 Reagents and Equipment Needed

▶ *Equipment*

- burette of 25 ml
- pipette of 50 ml
- beaker or Erlenmeyer (flask) of 250 ml
- agitator made of glass or magnetic stirrer
- measuring pipette of 5 ml

▶ *Reagents*

- normal caustic soda
- murexide
- EDTA 2 Na solution 0.01 M

3.3.4.3 Procedure

Pipette in a beaker or Erlenmeyer (flask) of 250 ml, exactly 50 ml of water to be analyzed; add 2 ml of normal caustic soda; add a trace of murexide; titrate with the EDTA M/100 until the colour change from salmon pink to purple.

3.3.4.4 Expression of Results

Calcium TH = calcium hardness (the same calculations as the total hardness).

3.3.4.5 Magnesium Hardness

It is the difference between the total hardness and the calcium hardness.

B – Water Treatment

1

Generalities

In general, natural water that we have at our disposal is not ready for a desired end-use, thus, its treatment is necessary.

In most cases, three phases are distinguished in water treatment.

1° *Before utilization*: we need to provide water with the characteristics required for its use: it is the pretreatment or external treatment.

The external treatment requires the use of equipments, filters, clarifiers, deaerators, ion exchange systems, water softeners, etc., the equipments having the purpose of reducing hardness and alkalinity, removing dissolved gases and suspended solids.

As we can see it from this short description, the pretreatment is mainly related to removal; it removes from water the impurities which would be harmful to the utilization process.

The pretreatment can be all the more meticulous since the conditions of use are more delicate, from single-use/purpose simple rough filtration for cooling water to a quasi-total demineralization for producing high pressure vapour.

2° *During utilization:* a treatment using additive is constantly practiced in order to keep water within the limit of the characteristics required.

This treatment is called water conditioning.

It has three main objectives, which are the following:

- to restrict the corrosion of the material to an acceptable level (< 50μ/year for steel);
- to maintain the level of thermal exchange, that is, to avoid scaling and formation of inorganic and organic sludges;
- to eliminate the bacteria in general and particularly the bacteria of the type legionella that cause the legionnaires' disease, which can be lethal.

3° *After utilization:* from then on, the water is qualified as wastewater. Its discharge into the natural environment can be made only by respecting the legislation in force.

Supplementary treatments can become indispensable.

Treatments realized belong to two large groups: the physical-chemical treatment and the biological treatment.

WATER TREATMENT

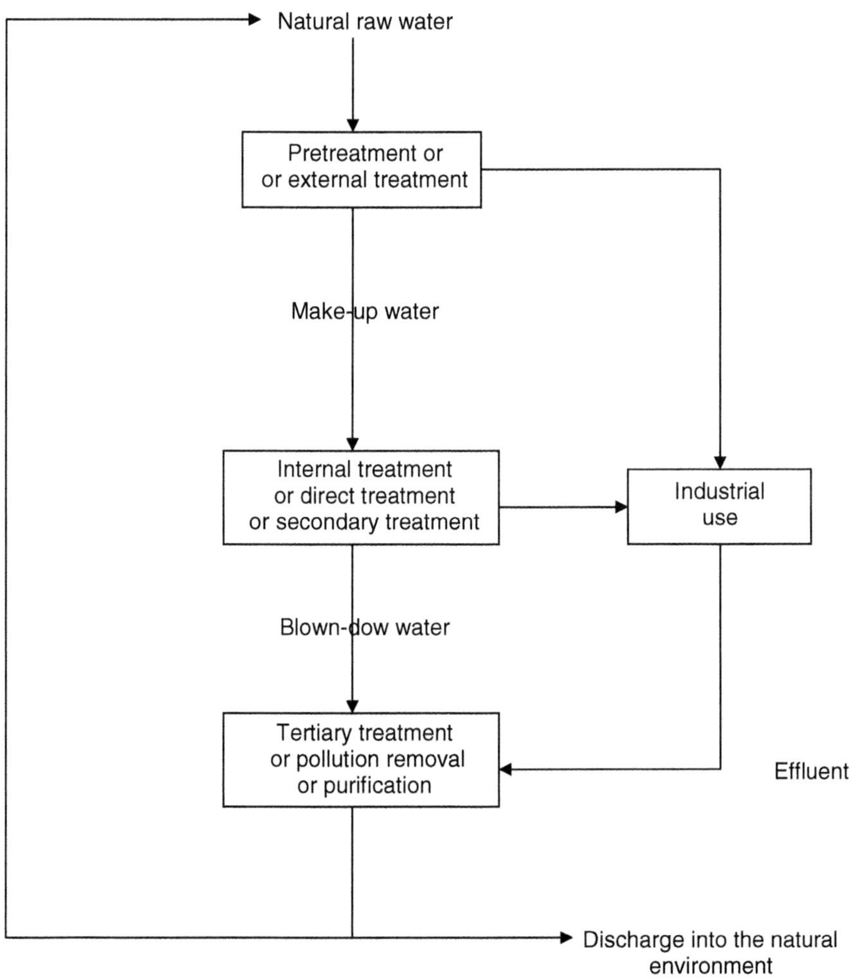

2

Pretreatment

Raw water used by industries is, in general, issued from one of the following two sources: runoff water and underground water. The runoff water includes water of rivers and lakes, and the underground water comes from wells or drillings/boreholes.

Other sources consist of effluents of the sewage plants and recycled water issued from industries.

In general, the underground water is harder, contains more alcalinity and dissolved solids in total than the runoff water. However, the underground water have, in general, more regularity as to their quality throughout the year, they are less turbid and are less sensitive to the microbiological contamination than the surface water.

The surface water has the enviable advantage of being immediately available, and it contains lower concentration of dissolved solids, in general.

In some cases, it is desirable to use a mixture of these two sources of water.

Undesirable impurities of the raw water are those caused by scale deposits and problems of fouling and clogging in the cooling system. They comprise:

1. suspended solids;
2. calcium;
3. silica;
4. alkalinity;
5. phosphates;
6. iron;
7. manganese;
8. organic matters;
9. dissolved gaz.

2.1 Coarse Screening or Screening

It is a matter of the inlet structures, which consists of all kinds of devices, from the screen bars spaced out from each other by some centimetres, in order to move pieces of wood and floating waste away, until the sieves with mesh opening of somme millimetres, intended to keep away the floating masses of algae and aquatic plants, which grow in the tanks/reservoirs.

2.2 Aeration or Oxidation by Air

In the pretreatment, the aeration is used for removing undesirable gases from water, namely, the carbon dioxide and hydrogen sulphide, both of which are corrosive. The aeration also intervenes in the removal of iron and manganese, combined in this case with a chemical treatment.

In the treatment of drinking water, the aeration is used for eliminating certain tastes and certain odours.

In the case of wastewater, the aeration provides the oxygen necessary for the biochemical oxidation of organic matters.

2.2.1 Principle of Aeration

It is a process which consists in thoroughly mixing water and air so as to reach or to approcah the state of equilibrium between the dissolved gases and the atmosphere.

The gas removal by aeration follows the Henry's law, which says that the solubility of a gas is directly proportional to its partial pressure in the atmosphere in contact.

At 20 °C, water in equilibrium with the atmosphere contains about:

— 15.8 ppm of nitrogen;
— 9.4 ppm of oxygen;
— 0.5 ppm of CO_2;
— only some traces of SO_2.

On the other hand, at saturation, water contains, at 20 °C:

— 1,700 ppm of CO_2;
— 3,900 ppm de H_2S.

As the partial pressure of the two gases is practically nil in the atmosphere, the complete removal of the gases can be obtained in the case where the pH is sufficiently low, in order there is no presence of bicarbonates or sulfhydrates, according to the following equilibrium:

$$CO_2 + H_2O \leftrightarrows HCO_3^- + H^+ \leftrightarrows CO^{--} + 2H^+,$$

equilibrium which has shifted to the left with the increase of the concentration of ions H^+, that is to say, with a decrease of pH.

2.2.2 Equipment Used

Two types of equipent are used:

2.2.2.1 Device with Waterfall

There are many variants based on this principle:

— free fall of water in droplets;
— fall of water on the successive plates in order to increase the contact;
— nozzles of water injection in order reduce the diameter of the drops.

We often use the plates covered with activated carbon (which has a catalytic action for the removal of iron and manganese) in order to promote the renewal of air; in backflow, we also use the forced draught aerators that remind impingement-plate columns (scrubbing towers?) or cooling towers.

2.2.2.2 Device for Diffused Air Aeration

The air diffusion is realized by forcing air to pass in the water, using submerged, perforated tubes, porous plates, etc.

This type of aeration is theoretically superior to the one using waterfall, in terms of obtaining a better surface of contact, as well as the advantage of having the air bubble rising velocity lower than the water drop falling velocity, but the cost of this equipment is higher than that of the equipment using waterfall.

2.2.3 Applications of Aeration

2.2.3.1 Carbon Dioxide Removal

In many cases, the aeration is used for removing the CO_2 liberated by a chemical treatment; for instance, after a treatment on resin in cycle Na, water is acidified before the aeration.

The necessity of acidification is clearly emphasized in the table 6 of page 59, which indicates, depending on the pH, the fraction of the total CO_2 present in the form of free acid, bicarbonate and carbonate.

2.2.3.2 Hydrogen Sulphide Removal

The hydrogen sulphide can be of purely chemical or biological origin, however, in all the cases, it is necessary to remove it, either because of its odour, or because of the corrosion that is provokes as in the case of the carbon dioxide; the removal can be complete only if the aeration is carried out in acid pH. The following table shows it as well.

Percentage of total sulphides present in the form of H_2S:

pH	% H_2S
5.0	98
6.0	83
6.5	61
7.0	33
7.5	14
8.0	4.8
9.2	0.32

2.2.3.3 *Iron Removal*

In general, iron is present in water in the form of ferrous bicarbonate. If we content ourselves with increasing pH of the water, we would obtain only an incomplete precipitation of iron in the form of ferrous hydroxide; it is thus necessary to oxidize the ferrous iron into ferric iron through aeration: the precipitate obtained will then be the ferric hydroxide hardly soluble.

$$4\ Fe(HCO_3)_2 + O_2 + 2\ H_2O \rightarrow 4\ Fe(OH)_3 + 8\ CO_2$$

As this operation is very sensitive to the pH, it is necessary to have a pH > 7.5, in order to obtain a correct removal of iron.

The ferric hydroxide is then removed through settling or filtration with or without coagulation.

2.2.3.4 *Manganese Removal*

The principle is the same as in the case of iron removal, and manganese is generally removed in the same time with iron; let us note, however, that it requires a slightly higher pH and a longer aeration.

Manganese must be removed, taking into account very low solubility of the manganese oxide (MnO_2), which is formed slowly and spontaneously in water containing dissolved oxygen.

2.3 Chlorination

The chlorination is a commonly used treatment, especially in the production of drinking water, in order to ensure its disinfection; in addition to its bactericidal effect, the chlorination has many advantages:

— colour and odour elimination;
— iron and manganese oxidation;

— oxidation of sulphides into sulphate;
— destruction of organic matters;
— oxidation of nitrites;
— bactericidal action.

During the clarification of water, we practice the coagulation in the same time as the prechlorination. It improves the coagulation.

2.3.1 Chlorination Chemistry

Chlorine is hydrolyzed in aqueous solution, liberating hypochlorous acid which is an active element.

$$Cl_2 + H_2O \rightarrow HClO + HCl$$

Below the pH of 6.5, the hypochlorous acid is present completely in the form of acid, while at pH > 9, it is completely ionized in the form of $C1O^-$, and it is the form HC1O that is by far the most active one in bactericidal terms.

This remark explains the relative efficiency of the chlorination of water whose pH is alkaline (in particular, superior to 8).

2.3.2 Other Chlorination Agents

Sodium hypochlorite and calcium hypochlorite are used for the same puprose as chlorine and they have the same action at the same pH, since we have the same active chemical entity.

2.4 Sedimentation or Settling

Surface water is usually pumped from a river to a reservoir. If we retain this water, the suspended solids that it contains can form a deposit. The velocity at which the particles settle depends on the size of these particles, their density and the viscosity of the water. The settling velocity can be calculated using the Stokes' law:

$$V = \frac{1/18\, g\, D^2 (d_1 - d_2)}{\eta}$$

where V: settling velocity of the particle in m/s;
D: its diameter in m;
d_1: density of the particle in kg/m^3
d_2: density of the liquid in kg/m^3
η: dynamic viscosity of the liquid in Pascal.

The typical velocities at which solid particles settle are the following, for a specific weight of 2.65 in water, and at the temperature of 50 °C :

Diameter of the particles mm	Order of size	Time necessary for forming a deposit layer of 30 cm
1.0	coarse sand	3 seconds
0.1	fine sand	38 seconds
0.01	coarse silt	33 minutes
0.001	Bacteria	35 hours
0.0001	Sludge (bottom mud)	230 days
0.00001	Colloids	63 years

The Stokes' law allows calculating the threshold/limit settling velocity of the particles that can be assimilated to spheres, in a liquid having a Reynolds number (N) inferior to 0.3.

2.4.1 Limitation of the Settling Method

Simple settling, like the one we have described, is usable for waters charged with particles having high enough falling velocity, that is to say, not too fine and of high density, which limits the method use to simple cases: natural water charged with slit, sand, even clay, for instance. In the case where the particles can not spontaneously fall at a sufficient velocity, we add the products/chemicals allowing increase this velocity.

2.4.2 Equipment Used

For small units, a simple tank or reservoir is used. Water is kept in the settling tank in a relatively sufficient restful state. Once deposits sink to the bottom, we use the upper layer of water.

With larger units, a continuous process is used. The size of the equipment is sufficient, so that the residual water velocity does not disturb the settling or sedimentation process.

The sludge removal is carried out on a continuous basis for larger plants, either using the accumulation of the sludge on an inclined surface and blow-down at the low point, or with flat or almost flat base, equipped with a sludge scraper device.

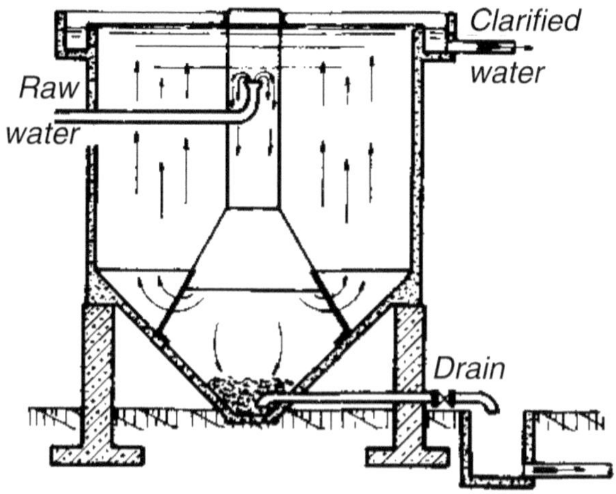

2.5 Clarification - Coagulation – Flocculation

2.5.1 Definition of the Process

In many cases, the sedimentation is not sufficient for removing suspended solids, as their falling velocity is too low. Therefore, we add coagulating or flocculatig agents, which bind the particles in elements of sufficient size in order they have a sufficient settling velocity.

Whatever the process of coagulation, the quality of raw water has noticeable effects on the performance of the chemical products used. A high turbidity increases the charge of solids in the clarifier, whereas a low concentration of solids does not provide enough cores capable of forming a floc. A high concentration of organic matters, the presence of detergents or other natural or synthetic dispersants tend to stabilize the colloidal suspension, thus requiring larger quantities of chemical coagulants. An alkalinity with low bicarbonate rate may necessitate the use of lime. A contamination by microbiological organisms may cause fouling or clogging of the material, as well as a carry-over of the floc, due to the gas formation.

It is therefore important to choose chemicals appropriate to the treatment of the relevant water. These chemicals and the doses to be used may vary depending on the variations of the quality of the water to be treated. It may be useful to implement beforehand a homogenization tank, in order to cope with brutal variations in the quality of the make-up water.

2.5.2 Coagulation Theory

The precipitation of a coagulant of the type of aluminium or iron salt, produces finely divided particles of the relevant hydroxide. The particles are charged positively, thus repel one another.

The neutralization of the positive charges by multiply negative ions such as the chlorides or sulphates provokes the coalescence between the particles, thus increasing their size, with appearance of a bulky precipitate.

During flocculation of hydroxides, the fine particles, which reduce limpidity, are found trapped in the flocculate, and thus eliminated from the liquid phase.

For each coagulant, there is an optimum pH range for obtaining a good coagulation, the pH which corresponds to a minimum solubility of the floc. This pH may depend on the mineralization of the water to be treated.

In general, coagulation requires:

— presence of a minimum quantity of aluminium or iron ions, in order to obtain an insoluble floc;
— presence of a strong anion such as sulphate or chloride;
— control of the pH within a precisely determined range;
— addition of a flocculant which increases the growth of the floc and its falling velocity.

Continuous measurement of the zeta potential may be used in order to adjust the amount of the coagulant depending on the quality of the water and to adapt it to the variations thereof.

2.5.3 Coagulants Used

Coagulants can be either organic or inorganic.

The most common inorganic coagulants are the following:

Alum: $Al_2(SO_4)_3$, 18 H_2O

The reactions involving the alum are the following:

(1) $Al_2(SO_4)_3, 18\,H_2O + 3H_2O \rightarrow 2\,Al(OH)_3 + 3\,H_2SO_4 + 14\,H_2O$

(2) $2\,Na\,AlO_2 + Ca(H\,CO_3)_2 \rightarrow Ca(AlO_2)_2 + 2\,Na\,H\,CO_3$

$2\,Na\,AlO_2 + Ca(OH)_2 \rightarrow Ca(AlO_2)_2 + 2\,Na\,OH$

Ferric sulphate: $Fe_2(SO_4)_3$ gives rise to the following reactions:

(3) $Fe_2(SO_4)_3 + 3\,Ca(OH)_2 \rightarrow 3\,Ca\,SO_4 + 2\,Fe(OH)_3$

$Fe_2(SO_4)_3 + 3\,Ca(H\,CO_3)_2 \rightarrow 2\,Fe(OH)_3 + 3\,Ca\,SO_4 + 6\,CO_2$

Ferrous sulphate gives the following reactions:

(4) $4\,Fe\,SO_4, 7\,H_2O + 4\,Ca(HCO_3)_2 + 8\,Ca(OH)2 + O_2 \rightarrow$
$4\,Ca\,SO_4 + 8\,Ca\,CO_3 + 4\,Fe(OH)_3 + 34\,H_2O$

$4Fe\,SO_4, 7H_2O + 4Ca(OH)_2 + O_2 \rightarrow 4\,Ca\,SO_4 + 4\,Fe(OH)_3 + 26\,H_2O$

Ferric chloride gives the following:

(5) $2\,Fe\,Cl_3 + 3\,Ca(HCO_3)_2 \rightarrow 2\,Fe(OH)_3 + 3\,Ca\,Cl_2 + 6\,CO_2$

$2\,Fe\,Cl_3 + 3\,Ca(OH)_2 \rightarrow 2\,Fe(OH)_3 + 3\,Ca\,Cl_2$

Polymer aluminium oxychlorides have as formula $Al_n(OH)_mCl_{3n-m}$. Their molecules have the composition of an inorganic polymer, while the aluminium is in the form of a polyanion ion.

The coagulating efficiency has increased from 1.5 to 2.5 in relation to the aluminium salts, with addition of a quantity of anions considerably reduced in relation to the conventional coagulants.

The alum should apply to (the context of) a pH of 6 to 7.

The ferric chloride may be applied to a pH situated between 5 and 11.

Other coagulants are used sometimes. These are:

— sodium aluminate $NaAlO_2$, highly akcalizing;
— ferrous sulphate $FeSO_4$, H_2O, which gives rise, in the absence of oxygen, to a precipitation of ferrous hydroxide more soluble than Fe $(OH)_3$.

Organic coagulants are the compounds of the type polymer, such as the polyamines and polydadmac having numerous positively charged sites. Resins, formaldehyde melamine and dicyandiamide are commercially available as well. Their use has been expanding.

They have an advantage of restricitng definitely the quantity of sludge obtained after settling. From the purely economic point of view, their use is to be favoured, given the cost of the sludge draining and treatment, when discharge is considered to be necessary, given their composition.

2.5.4 Use of Organic Flocculants

Synthetic polyelectrolytes have been more and more in use, in order to increase the efficiency of or replacing the inorganic coagulants.

They are constituted by polymers of high molecular weight and soluble in water, with ionization of their functional groups. They may be:

— anionic;
— cationic;
— non-ionic.

Their operations are different depending on their polarity.

▶ *Flocculation through charge neutralization*

The cationic polyelectrolytes, through their quaternary ammonium compound groups, provoke the electric discharge of colloids, by adsorbing to the solid –liquid interface, according to a reaction very close to that involved in the coagulation with inorganic/mineral electrolytes.

This ionic neutralization considerably increases the adhesion of adjacent particles (action connected to the nullifying of the Zeta potential).

▶ *Floculation through physical-chemical adsorption of polymers*

Anionic and non-ionic polymers provoke flocculation through a phenomenon of adsorption connected to the formation of hydrogen bonding (the case of $CONH_2$ radicals or polyacrylamide).

The phase of adsorption is followed by the formation of bridges between the colloids, which constitute a cross-linking, with formation of a floc.

The phase of cross-linking is greatly helped by the polymer's molecular weight.

► *Choice of the coagulant/flocculant system*

Important parameters to be considered are the following:

- turbidity of the water to be treated;
- concentration of organic matters;
- bicarbonate alkalinity.

A high turbidity will lead to a significant volume of sludge to discharge, whereas a low turbidity will produce too few and too small flocs.

A high concentration of organic matters will stabilize the colloidal matters and will necessitate too high dosage rate of coagulant.

The treatment with coagulation and flocculation is carried out in two stages:

a) the coagulant is first added to the water to be treated in a system equipped with a vigorous mixing (device);
b) the flocculant is then added to the water in a system equipped with a device for slow mixing, in order to promote the formation of voluminous flocs and gather all the particles present.

The water thus treated passes finally through a clarifier/settling tank, from which the sludge is removed through the bottom. The treated water is collected by overflow.

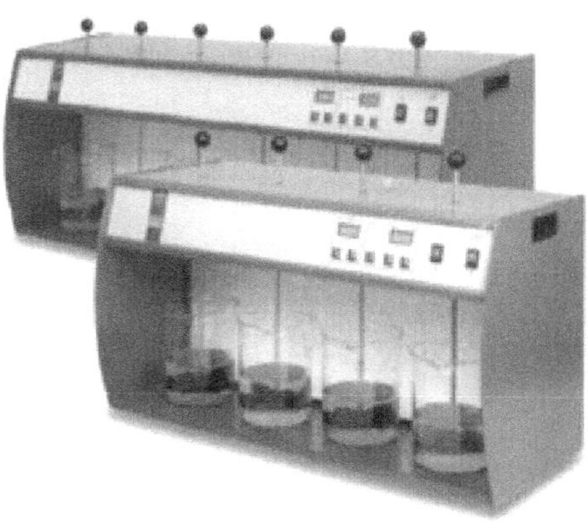

The most widespread method for searching an efficient program of coagulation/flocculation is to carry out, in laboratory, comparative tests on different products, using a jar test like the one that we see on the above-presented photograph.

We can make vary, as much as we like, the stirring speed so as to quickly obtain a mixture, the growth of the floc with slow agitation, then measure the settling velocity.

This way of proceeding allows determining:

a) the effects of the coagulants, individually and in combination, on the water which is examined;
b) approximate dosage rates of chemicals;
c) the effects of different speeds of agitation;
d) the pH range for the most efficient flocculation;
e) the order and the most appropriate moment for adding the chemicals.

2.5.5 Equipment Used

The commonly used equipment consists of a settling tank/clarifier with sludge blanket or a settling tank/clarifier with solid particles contact.

The settling tank with sludge blanket, illustrated in the figure presented below, is a compact unit. The raw water, which is injected into it, is treated with chemicals in the primary zone and, almost immediately, it is brought into contact with the sludge particles, which contribute to the development of flocs.

The water which passes at the bottom of the cone losses the most heavy floc particles, thanks the gravitational force. The lighter particles, which tend to go upwards again with the water, are trapped by the sludge blanket, which results in promoting the floc seeding and removal through absorption and filtration by means of the sludge layer.

Retention time is usually 60 to 90 minutes.

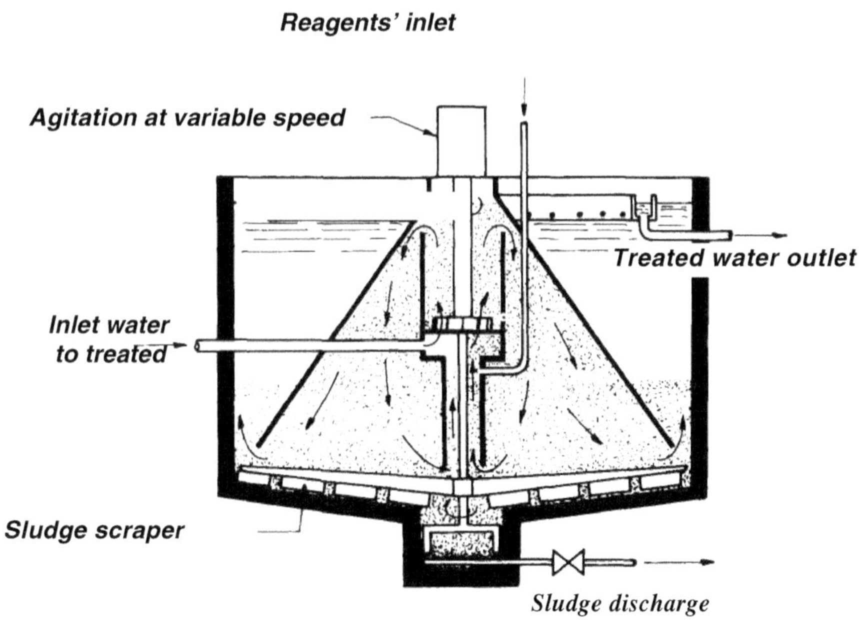

SLUDGE BLANKET CLARIFIER

The solids contact clarifier will be described in the chapter referring to the process of lime softening and the softening using caustic soda.

Some points should be examined:

- the choice of the best points for adding the flocculants and coagulants;
- the speed of mixers and agitators may have a significant effect on the quality of the water produced;
- sludge should be removed regularly, otherwise, it forms a compact mass difficult to remove.

2.6 Lime Softening

2.6.1 Definition of the Process

This process aims at removing the ions Ca^{++} and Mg^{++}, which contribute to the water hardness, by precipitating them through the combined action of lime and sodium carbonate. This operation has the consequence of not only reducing considerably the hardness, but also has an advantage of removing the dissolved CO_2, the turbidity, as well as reducing the concentration of silica, iron and manganese.

The only process which is actually still in use is the low temperature process where the water is treated at its inlet temperature.

The following chemical reactions occur:

$$Ca(HCO_3)_2 + Ca(OH)_2\ Mg \rightarrow 2\ CaCO_3 + 2\ H_2O$$
$$(HCO_3)_2 + Ca(OH)_2 \rightarrow MgCO_3 + CaCO_3 + 2H_2O$$

$$Mg(HCO_3)_2 + Ca(OH)_2 \rightarrow Mg(OH)_2 + CaCO3$$
$$MgCO_3 + Ca(OH)_2 \rightarrow Mg(OH)_2 + CaCl_2$$
$$MgCl_2 + Ca(OH)_2 \rightarrow Mg(OH)_2 + CaSO_4$$
$$NaHCO_3 + Ca(OH)_2 \rightarrow CaCO_3 + NaOH + H_2O$$

These reactions show that all the ions of Ca and Mg are removed (within the limits of their solubility) in the form of $CaCO_3$ and Mg $(OH)_2$, whereas the residual salts are chlorides and sulfates of sodium and possibly that of caustic soda.

Reduction of the silica concentration is the result of the absorption of the soluble silica by the magnesium hydroxyde. The more magnesium hydroxyde is produced, the more silica is reduced. If there is not enough magnesium of natural origin in the water, we can add it in the form of dolomitic quicklime or of magnesium oxide. The following figure shows this relation.

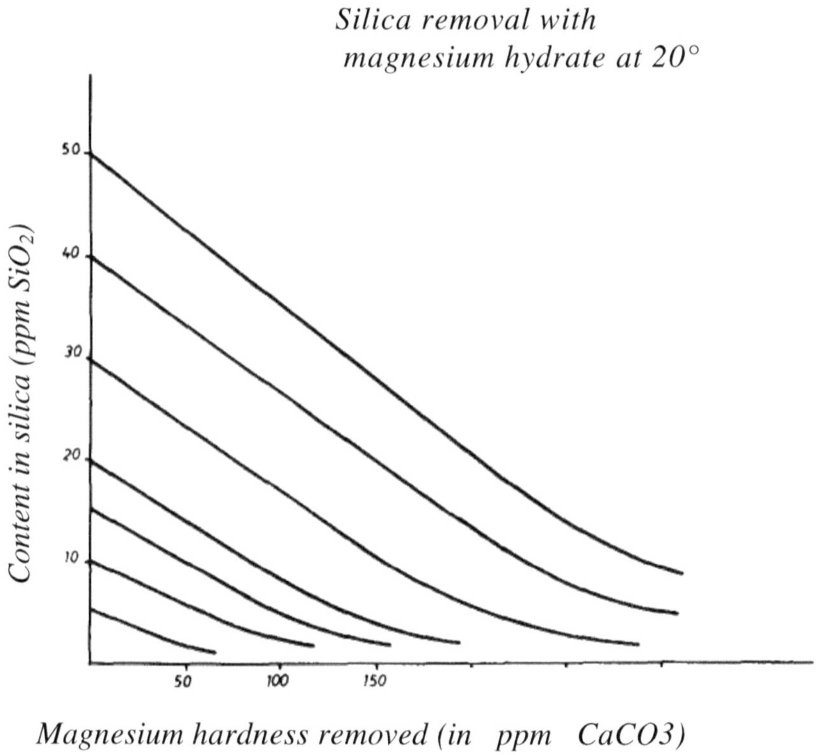

An estimation of the needs of lime may be carried out:

Weight of lime 100 % in kg for 10 m^3 of water = CO_2 + 0,44 MA + 0.019 Mg
Concentration of free CO_2 in ppm
MA = total alkalinity expressed in ppm of $CaCO_3$
Mg = ppm of magnesium expressed in Mg.
Concentration of CO_2 is read on the diagram of the page 21.
Let us remind that the carbonate hardness is equal to:
total hardness — total alkalinity (TH — MA).

2.6.2 Equipment Used

It consists of solid-liquid contact clarifiers so as to ensure a correct growth of the crystals before the settling stage.

Thus, the water is brought into tight contact with the crystals' suspension. The cheimical reactants are added at this moment and the precipitation occurs quickly, while the sludge particles serve as the crystallization cores. The solids thus formed are of such size and weight that they can quickly be removed.

Usually, there is a clear line of separation between the suspension and the clear water.

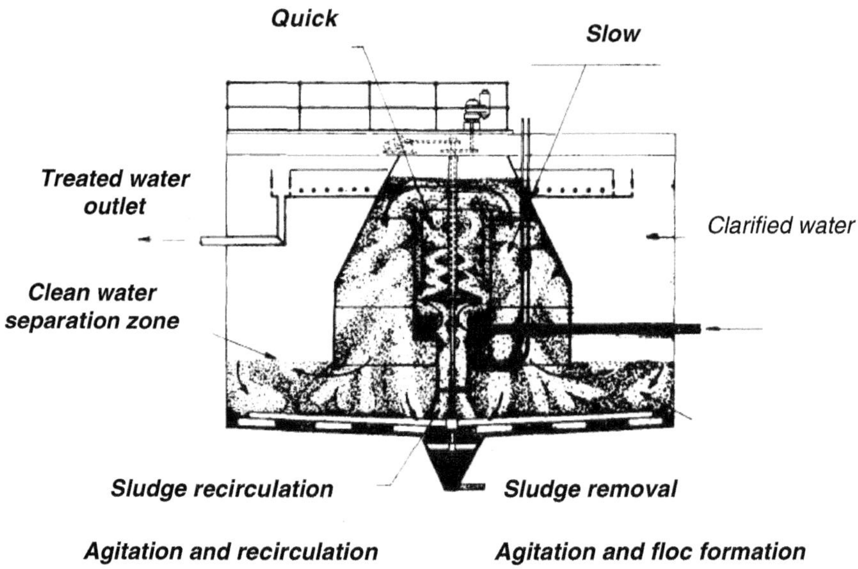

Fig. Sludge contact lime softening

These units operate with upward overflow velocity of 3 to 6 cm/minute for a retention time of about one hour.

The use of coagulants is useful, in general.

A well-designed unit removes all the suspended solids, reduces the calcium hardness until 2 – 3.5 °F, reduces the total alkalinity until 3 — 5 °F and removes phosphates, iron and manganese.

2.7 Softening Using Ion Exchange Resins

2.7.1 Ion Exchange through Resins in Sodium Cycle

2.7.1.1 Principle

Principle: Passage of the water on strong cationic ion exchange resins regenerated in the form of sodium. In general, a correctly-operating softener must deliver water having a total residual hardness (TH) inferior to 0.5°f.

$$R\text{-}Na_2 + Ca^{++} \rightarrow CaR + 2\ Na+$$

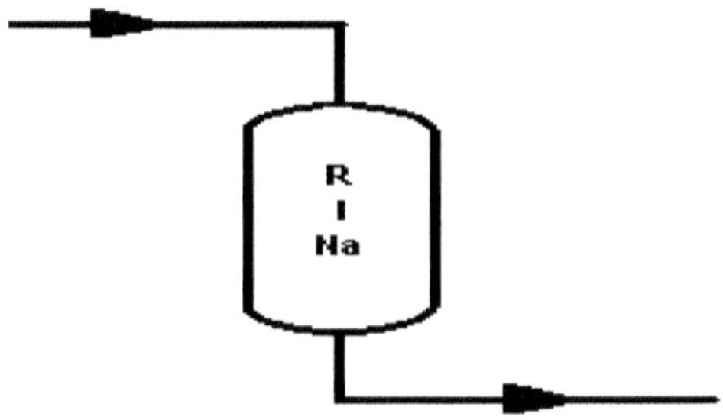

The ion exchanger molecule is regenerated with a concentrated solution of salt, according to the equation:

$$\text{Ca R} + 2\ \text{NaCl} \rightarrow \text{Na}_2\text{R} + \text{CaCl}_2$$

The reaction is governed by the fundamental law of mass action. In the softening cycle, the calcium ion prevails and it is the first reaction that occurs. In the regeneration cycle, the sodium ion or hydrogen ion prevails, and it is the second reaction that occurs. Given the fact that the regeneration necessitates the presence of an excess of salt, a given extent of hardness can not be removed with the same amount of salt consumed.

The consumption of salt adds up to about 2.3 — 3.5 kg of NaCl per kg of $CaCO_3$ removed.

The water which supplies the ion exchange units should not contain suspended solides. Only calcium and magnesium are removed and the total percentage of dissolved solids remains the same. For this reason, the process is rarely used for the make-up water of cooling towers, and when it is used there, only a part of the water is needed to be treated, since the effluent's hardness is close to zero.

It is to be noted that the use of resins in the sodium cycle does not modify the alkalinity of the water, since the bicarbonates of Ca and Mg are transformed in sodium bicarbonate.

► *Softening carbonate removal*

Principle: Passage of the water on weak cationic ion exchange resins regenerated in the form of H^+, then on a softening resin. In general, a softening carbonate removal, when operating correctly, must deliver water having a residual total hardness (TH) inferior to 0.5°F, a pH situated between 4.5 and 5 and a MA (m-alk) inferior to 1°F.

In this type of facilities, resins widely known as carboxylic resins, which ensure the carbonate removal, transform the bicarbonates into carbon dioxide. The carbon dioxide is removed in a degassing tower.

The non-carbonate hardness generates the formation of acids corresponding to the associated anions of either hydrochloric acid or sulfuric or nitric acid.

The free acidity generated by the concentration of the salts of strong acids (SSA) must be neutralized for certains cases of use.

2.7.1.2 Equipment

The equipment consists of tanks capable of resisting to the water circuit pressure, with additional tanks necessary for the supply of the product of regeneration.

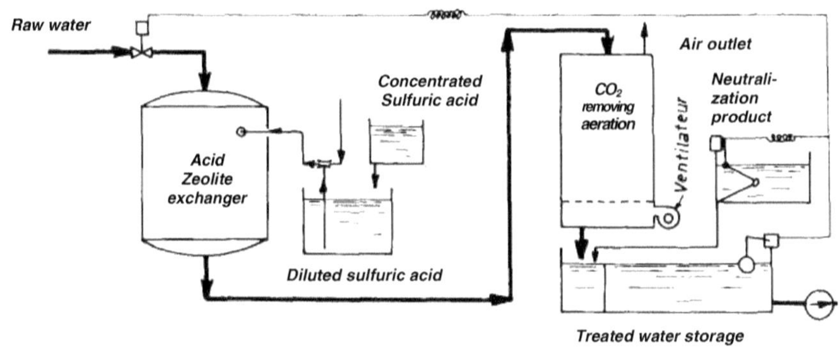

SOFTENING USING ACID ZEOLITE AND NEUTRALIZATION

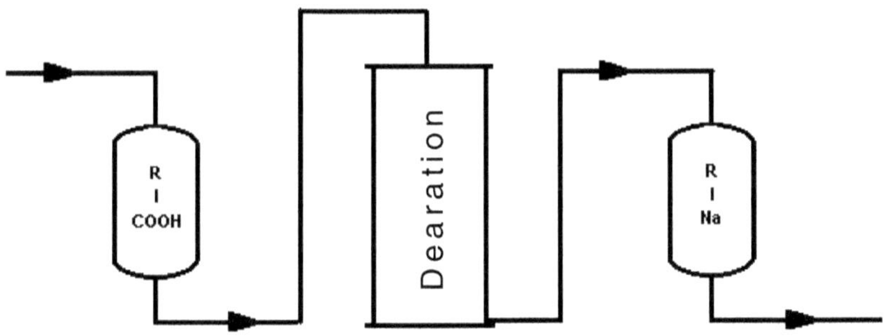

2.7.2 Complete Demineralization

Ion exchange resins consist of synthetic polymers made insoluble through cross-linking (of phenolic or acrylic nature, in general) and carrying ionisable moieties.

Depending on the nature of the moieties, we distinguish the following types of resin:

— cation exchange resins, which carry anionic groups, generally carboxylic R — COOH (weak acid exchangers), or sulphonic R — SO_3H (strong acid exchangers);
— anion exchange resins, which carry cationic groups, either amino, primary, secondary or tertiary (we have then weak base exchange resins), or quaternary ammonium compounds (strong base resins).

R — NH_2 primary amino group

R — NH secondary amino group

CH_3

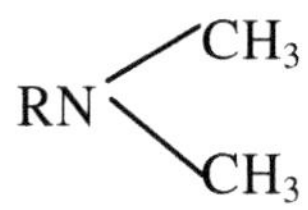

tertiary amino group

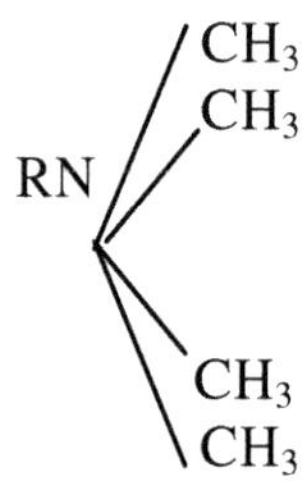

quaternary ammonium group

2.7.3 Use of the Resins

The total demineralization of water necessitates the use of two types of resin.

The acid resin exchanges the cations, the base anion exchange resin then removes the anions left in the water already decationized.

The weak base exchanger resins remove the strong anions (sulphates, chlorides, nitrates). On the other hand, the weak anions (silicates and carbonates or bicarbonates) are removed by the strong base resins.

The water going out from the plant has finally lost all its ions, and the ions H^+ and OH^-, left by the two resins are recombined to form water.

2.7.4 Equipment Used

The simplest plant (see the figure below) consists of only two columns of cation and anion exchangers set up one after the other. This device can be used only for the treatment of water having low bicarbonate hardness (where the release of CO_2 is low).

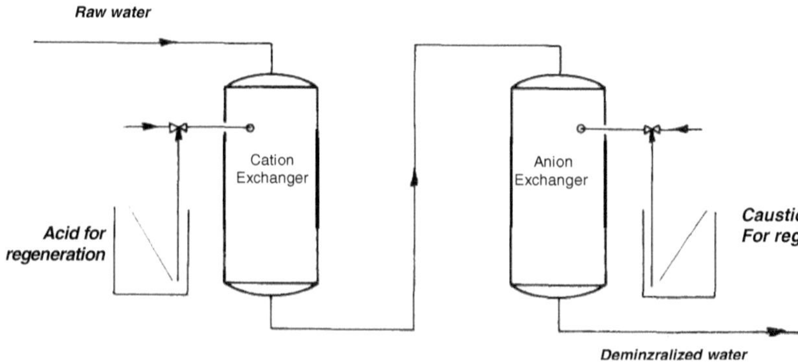

DEMINERALIZATION WITH 2 SEPARATE BEDS

As in the case of the softening carbonate removal, it is necessary to consider a degassing column (removal of carbon dioxide between the strong anion and the strong cation).

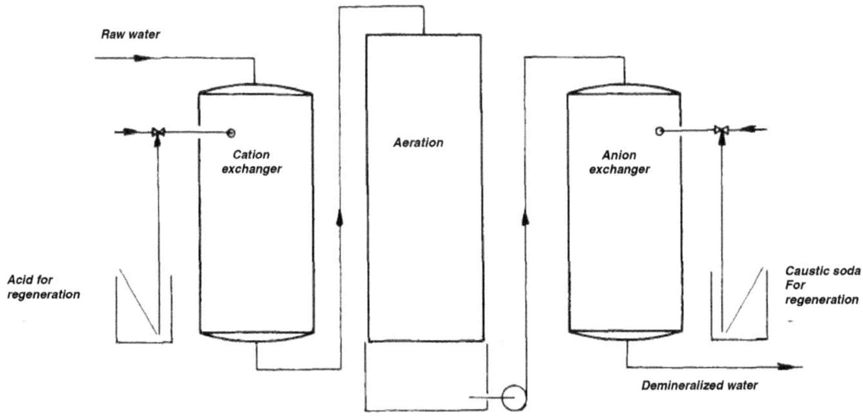

Fig. Demineralization with 2 beds and a deaerator

2.8 Comparing Different Treatments

(see table 7, page 60)

We have indicated, for the same raw water, effects of different treatments feasible in pretreatment and allowing, depending on the purpose concerned, see immediately what solution to choose.

2.9 Filtration

When we need perfectly limpid water, whether it be raw water or pretreated water, it is rare that the flocculation or softening using lime leaves the water completely free of suspended matters, thus a filtration proved to be necessary.

2.9.1 Principle

Filtration consists in making percolate a liquid through a porous medium so as to screen out solid particles, which remain there after sedimentation carried out correctly; after chemical treatment, most of the heavy particles duly coagulated (must) have been removed. It is only the finest particles, suspended solids that reach the filter.

As the solid particles reach the filter layer, they settle more or less uniformly on the surface and stop the particles arriving subsequently. The efficiency of the filter thus increases depending on its clogging. In practice, the penetration of the filtering material does not exceed some centimetres and, at the moment, the porosity of the filter, which has only been decreasing, becomes too weak to continue using it (too significant head loss). A countercurrent washing is then used in order to remove the particles accumulated with the filter layer.

2.9.2 Nature of the Filtering Components

In general, they consist of several successive layers of sand or silex having different grain sizes. The grain sizes are carefully selected in order to ensure that the progressive clogging of the filter is made at the highest possible level, whereas the total layer of the filtering components is 30 to 70 cm.

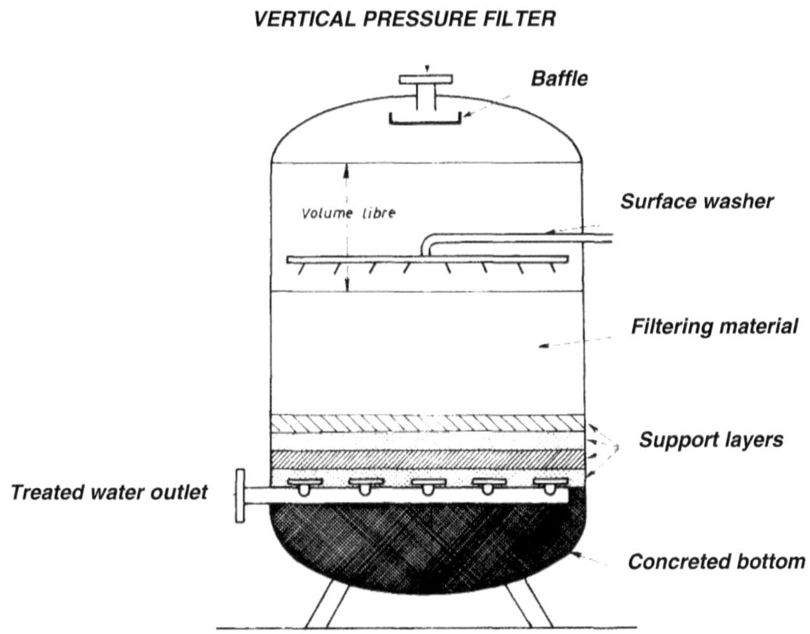

2.9.3 Type of Filters

Filters can be of the type pressure filter or gravity filter. The filter rate is, in general, of 100 litres per m^2 of filtering surface area per minute. The countercurrent washing rate varies from 300 to 400 litres per m^2 per minute, as far as the the sand filters are concerned, and from 200 to 300 litres per m^2 per minute, as far as the anthracite filters are concerned.

Suspended solids may sometimes be removed from raw water through filtration, in particular, when the filter is treated with a cationic polyelectrolyte, after each countercurrent washing. The filtration may or may not be necessary after the clarification process, depending on the concentration of suspended solids tolerable for the make-up water.

In general, filtration is used after the softening using lime and caustic soda, and the suspended solids should be removed from water before the ion exchange.

Filtration of a part of the recirculated flux of a cooling circuit allows removing most of the suspended solids present and limiting the problems of fouling and deposits, which are the base for the development of biofilms.

In general, filtration of 5 to 10% of the recycled throughput (of water) is sufficient. By removing a part of the sources of deposit, the filtration contributes to the prevention of the risk of legionella, to the reduction of the costs of treatment and to make the performance reliable.

The dimensioning of the filter design depends on the composition and the particle size of the suspended solids to be removed, on the throughput of the water to be treated and of course on the desired result expressed in mg/l of suspended solid or in turbidity.

Certain techniques allow meticulous filtration, which are similar to desalination, because they allow the ion filtration:

- microfiltration;
- ultrafiltration;
- reverse osmosis.

They are presented in the following chapter.

2.10 Membrane Techniques

Processes of separation by membranes have seen, over the latest years, a considerable development in the field of water treatment, thanks to the use of new synthetic membranes.

A membrane is a material which has the property of exhibiting a selective resistance to the transfer of different constituents of a fluid, thus allowing separation of some elements constituting the fluid.

2.10.1 Microfiltration and Ultrafiltration

Among all the processes using membranes, the tangential microfiltration and the ultrafiltration are well developed in industries.

These two techniques are close to each other and can be classified depending on the size of particles or macromolecules that they separate.

The principle of separation is the difference of pressure on both sides of the membrane.

The filtration takes place most often in tangential mode, that is to say, the fluid circulates in parallel to the membrane, and this is in order to limit the accumulation of species stopped at the surface of the membrane, contrary to the filtration in dead-end mode.

The tangential microfiltration is distinguished from the ultrafiltration by:

- more significant size of the constituents retained/held;
- usually lower working pressure (Transmembranary pressure < 3 bars);
- filtration rate often more significant.

In theory, all what applies to the ultrafiltration applies to the microfiltration, both in terms of the equipment used and the influence of operational parameters and clogging phenomena.

The membranes are "regenerated" by backflushing, which often turns out to be insufficient, and should be completed by an appropriate chemical treatment.

In most of the applications of dead-end microfiltration, flat membranes used in laboratory, filtration cartridges on flat or pleated membranes, as well as filtering components are discarded, once clogged.

Microfiltration: filtering components (doc. CFM)

2.10.2 Reverse Osmosis

Osmosis is the transfer of a solvent (water, in most cases) through a membrane under the action of a concentration gradient.

Let us consider a system with two compartments separated by a selective permeable membrane, containing two solutions of different concentrations.

The phenomenon of osmosis will result in a flux of water from the dilute solution towards the concentrated solution.

If we try to stop this flux of water by applying a pressure on the concentrated solution, the quantity of water transferred by osmosis will decrease. There will be a moment where the pressure applied will be such that the flux of water will stop.

If, to simplify, we suppose that the dilute solution is pure water, this pressure of equilibrium is called osmotic pressure. An increase of the pressure beyond the osmotic pressure will result in a flux of water in the direction opposite to that of the osmotic flux, that is to say, from the concentrated solution towards the dilute solution: this is the phenomenon of reverse osmosis.

The reverse osmosis allows keeping all the dissolved salts, organic molecules of 10^{-7} mm and monovalent salts under a pressure inferior or equal to 80 bars.

Technical progress and market growth in the water treatment field make the membrane techniques increasingly competitive compared to the traditional treatment methods.

One example: more than 50% of the plants producing soft water through the desalination of sea water use the process of reverse osmosis.

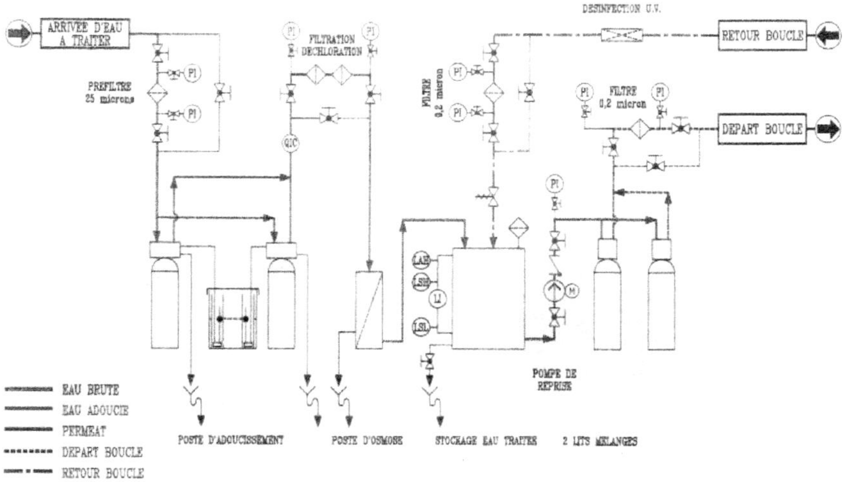

Plant for desalination of sea water using reverse osmosis (doc. CFM)

2.10.3 Problems Encountered and Their Solutions

It is necessary to avoid any deposit formation in the membrane modules. Otherwise, there will be a quick decrease in the flow rate of the water treated, leading to the interruption of the equipment operation.

There are two kinds of deposit risks /

- mineral deposit: calcium carbonate, calcium sulfate, barium sulfate and strontium sulfates, tri-calcium phosphate, silica in different forms.
- biological fouling (bio-fouling), explained by the presence of suspended solids in the make-up water.

Good use of installation leads to setting up the following:

- preventive treatments, to prevent the risks of mineral and biological deposits;
- curative treatments (cleaning), whose frequency will be such that the reversibility of the exchange properties will be assured.

a) Scale-formation risk:

All f the mineral salts presents in the make-up water concentrated in "concentrate", whose low rate is equal to the make-up flow minus the permeate flow (Flow of permeated water).

The concentration ratio is equal to the feed flow / concentrate flow.

This concentration ratio is theoretically determined by studying the possible maximum concentration, which is function of the solubility product of different insoluble salts present in the water. The possible maximum ratio is the one, for which the least soluble salt will remain in solution.

b) Fouling risk:

Fouling is caused by deposit of particles to the surface of the membranes, particles which are not evacuated by the countercurrent cleaning. All of the these particles constitute the "cake", which is compressible.

It has 3 different origins:

- colloids present in the make-up water (micro algae, clay, insoluble polymers, etc.);
- metallic oxides and hydroxides formed following uncontrolled pH variations;
- biological: formation of a bio-film constituted by microorganisms (bacteria, algae, fungi), got sticky in a matrix of polysaccharide secreted by these microorganisms.

2.10.3.1 Preventive Treatments

1) Filtration (pretreatment)

Filtration removes colloids.

2) Softening (pretreatment)

Softening removes calcium and eliminates the risk of calcium carbonate precipitation, which constitutes the most frequent scale-formation risk.

3) Chlorination

Such a treatment is realized with Javel water, aiming to have excess free chlorine of 0.1 mg/l.

In the case of polyamide membranes, dechlorination with sodium bisulphate is obligatory, in order to avoid the destruction of the membranes sensitive to chlorine.

4) Chemical scale preventives

The most frequently used method to avoid scale deposit consists in injecting, into the raw water, scale inhibitors. The main active materials used for this purpose are the following:

- polyphosphates;
- phosphonates;
- polyacrylates.

The choice of active material depends on the nature of the insoluble salt, of which we wish to avoid the precipitation, on the supply need of the permeated water, etc.

2.10.3.2 Curative Treatments

Despite the use of above-mentioned preventive treatments and controlled conditions of their use, progressive fouling of the membranes is still observed with time.

Chemical cleaning is necessary, in order to maintain the operation conditions of the installation close to its nominal conditions.

Curative cleaning products are the formulations which may be complex. These formulations are divided into two main categories: acid and basic.

The choice of a product is associated with the nature of fouling deposits present and identified:

If the deposit is of carbonate origin, then acid cleaning is to be considered.

If the deposit is of biological or colloidal origin, or based on amphoteric hydroxides, then alkaline cleaning will be effective.

These formulas can contain other active materials: sequestering (to dissolve calcium salts) wetting (to facilitate the penetration of the acid or base into the cake), emulsifying (to disperse greasy deposits), etc.

Implementation of a cleaning must always be preceded by an analysis of the cake, which considerably increases chances of success.

The frequency of cleaning is determined depending on experience and monitoring of the traversing flow. In general, cleaning must be carried out, when a decrease of 15% of the nominal flow is observed.

A decision of cleaning taken too late may prove ineffective and lead to the replacement of the membranes, as the elimination of the cake is impossible.

2.11 Desalination of Sea Water by Distillation

Production of drinking water using multiple-effect evaporator is still a commonly used technique, although it is the membrane techniques that have increasingly been used.

In this process, the sea water is brought to a temperature higher than the boiling point of water.

In order to have the highest possible conversion rate, the blow-down flow is limited to the maximum, which results in calcium sulfate over saturation, magnesium carbonate over saturation, etc., which would cause scale-formation on the heat exchange surfaces left without treatment.

The active materials allowing the highest conversion rate are the following:

- polymaleates;
- polyacrylates of low molecular weight;
- certain phosphonates.

The dosage of these formulas is between 1 and 3 ppm in relation to the make-up water.

Foaming phenomena occur sometimes, and the use of anti-foaming agent based on silicones brings an effective solution (some ppm).

Table 1. Constants of the Water Steam/Vaporization

Température °C	Pressure of saturated vapour kg /cm^2	Specific volume of the Liquid m^3/kg	Specific volume of the vapour mVkg	Mass enthalpy of the liquid kcal/kg	Mass enthalpy of the vapour in kcal/kg
0	0.00622	0.00100	206.4	0	595.4
50	0.1258	0.00101	12.6	49.95	618.5
100	0.0333	0.00104	1.674	100.04	639.2
150	4.854	0.00109	0.393	150.93	656.0
200	15.85	0.00116	0.127	203.6	667.5
250	40.60	0.00126	0.0501	259.5	670.0
300	87.7	0.00142	0.0218	321.8	657.5
350	168.7	0.00179	0.0089	403.7	617

Table 2. Mass of Water Vapour in Saturated Air (g/m^3)

Press. atm. relative Temp. °C	0.0	1.0	2.0	3.0	4.0	5.0	6.0	7.0	8.0	9.0
- 20	0.892	0.810	0.737	0.673	0.613	0.557	0.505	0.457	0.413	0.373
- 10	2.154	1.978	1.811	1.658	1.519	1.395	1.282	1.177	1.079	0.982
- 0	4.835	4.468	4.130	3.813	3.518	3.244	2.988	2.752	2.537	2.340
+ 0	4.835	5.176	5.538	5.922	6.330	6.761	7.219	7.703	8.215	8.757
10	9.330	9.935	10.574	11.249	11.961	12.712	13.505	14.339	15.218	16.144
+ 20	17.118	18.143	19.222	20.355	21.546	22.796	24.109	25.487	26.933	28.450
+ 30	30.039	31.704	33.449	35.275	37.187	39.187	41.279	43.465	45.751	48.138

Table 3. Weight of Saturated Water Vapour per KG of Dehumidified Air

Temperature °C	Weight of water vapour, in g, per kg of dehumidified air under total pressure (in mmHg) of		
	740	760	780
—20	0.783	0.763	0.743
—10	1.76	1.72	1.67
— 5	2.61	2.54	2.48
0	3.96	3.87	3.76
5	5.66	5.53	5.36
10	7.95	7.71	7.52
15	11.00	10.80	10.50
20	15.3	14.8	14.5
25	20.7	20.2	19.70
30	28.3	27.5	26.96
35	37.6	36.6	35.60
40	50.1	48.9	47.60
45	67.3	65.3	63.5
50	89.5	86.8	84.3
60	159.0	154.0	149.5
70	291.5	289.9	279.6
80	589.0	554.0	582.2
90	1557.0	1430.00	1317.7

Table 4. Weight of dehumidified air, weight of water vapour in the air at 75%, 50% and 25% of saturation

A = weight of air contained in 1 m^3 of air (in kg).
V = weight of vapour per kg of dehumidified air (in g).

Temperature °C	at 75 %		at 50 %		at 25 %	
	A kg	V g	A kg	V G	A kg	V g
— 20	1.390	0.575	1.390	0.38	1.391	0.200
— 10	1.339	0.900	1.340	0.57	1.341	0.447
0	1.286	2.880	1.289	2.20	1.290	1.000
10	1.237	5.770	1.240	3.84	1.243	1.920
20	1.184	11.030	1.190	7.37	1.197	3.690
30	1.128	20.500	1.141	13.50	1.153	6.680
40	1.067	35.100	1.087	23.50	1.107	11.500
50	0.994	62.700	1.026	40.60	1.060	19.700
60	0.899	109.000	0.952	68.90	1.004	32.700
70	0.789	189.000	0.867	114.90	0.946	52.400
80	0.665	317.000	0.763	196.80	0.882	82.700
90	0.458	700.000	0.628	340.70	0.799	134.000
100	0.176	2750.000	0.433	700.00	0.689	220.000

Table 5. Expression of concentration in degrees

	Formule	Molecular weight	Value ;n mg/l			
			Milli-equiva-lent/l	French degree	German degree	English degree
1° Salts of Ca and Mg						
Calcium carbonate	$CaCO_3$	100	50	10.0	17.8	14.3
Calcium bicarbonate	$Ca (HCO_3)_2$	162	81	16.2	28.9	23.1
Calcium sulphate	$CaSO_4$	136	68	13.6	24.3	19.4
Calcium chloride	$CaCl_2$	111	55.5	11.1	19.8	15.8
Quick lime	CaO	56	28	5.6	10.0	8.0
Hydrated lime	$Ca (OH)_2$	74	37	7.4	13.2	10.5
Magnesium carbonate	$MgCO_3$	84	42	8.4	15.0	12.0
Magnesium bicarbonate	$Mg (HCO_3)_2$	146	73	14,6	26.1	20.9
Magnesium sulphate	$MgSO_4$	120	60	12,0	21.4	17.1
Magnesium chloride	$MgCl_2$	95	47,5	9.5	17.0	13.5
Magnesia	MgO	40	20	4.0	7.1	5.7
	$Mg (OH)_2$	58	29	5.8	10.3	8.2
2° Anions						
Carbonate	CO_3^{--}	60	30	6.0	10.7	8.6
Bicarbonate	HCO_3^{--}	61	61	12.2	21.8	17.4
Sulphate	SO_4^{--}	96	48	9.6	17.3	13.7
Sulphite	SO_3^{--}	80	40	8.0	14.3	11.4
Chloride	Cl^-	35,5	35.5	7.1	12.7	10.2
Nitrate	NO_3^-	62	62	12,4	22.1	17.7
Nitrite	NO_2^-	46	46	9.7	16.4	13.1
Phosphate	PO_4^{---}	95	3131,66	6,32	11,2	9.03
Silicate	SiO_2	60	60	12.0	21.4	17.1
3° Acids						
Sulphuric	H_2SO_4	98	49	9.8	17.5	14
Hydrochloric	HCl	36,5	36.5	7.3	12.8	10.3
Nitric	HNO_3	63	63	12,6	22.5	18
Phosphoric	H_3PO_4	98	32,66	6,52	11,6	9.31
4° Cations and oxides						
Calcium	Ca^{++}	40	20	4.0	7,15	5.7
Magnesium	Mg^{++}	24,3	12.1	2.43	4,35	3.47
Sodium	Na^+	23	23	4.6	8.2	6.6
	Na_2O	62	31	6.2	11.1	8.8
Potassium	K^+	39	39	7.8	13.9	11.2
	K_2O	94	47.1	9.4	16.8	13.4
Iron	Fe^{++}	55,8	27.9	5.6	10.0	8.0
Aluminium	Al^{+++}	27	9	1.8	3.2	2.6
	Al_2O_3	102	17	3.4	6.1	4.85

Table 5. (*continued*)

	Formule	Molecular weight	Value in mg/l			
			Milli-equiva-lent/l	French degree	German degree	English degree
5° Bases						
Caustic soda	$NaOH$	40	40	8.0	14.3	11.4
Potash	KOH	56	56	11.2	20.0	16.0
Ammonia	NH_4OH	35	35	7.0	12.5	10
6° Various salts						
Sodium bicarbonate	$NaHCO_3$	84	84	16.8	30	24
Sodium carbonate	Na_2CO_3	106	53	10.6	18.9	15.1
Sodium sulphate	Na_2SO_4	142	71	14.2	25.3	20.3
Sodium chloride	$NaCl$	58.5	58.5	11.7	20.9	16.7
Sodium phosphate	Na_3PO_4	164	54.7	10.9	19.5	15.6
Sodium silicate	Na_2SiO_3	122	61	12.2	21.8	17.4
Potassium carbonate	K_2CO_3	138	69	13.8	24.6	19.7
Potassium bicarbonate	$KHCO_3$	100	100	20	35.7	38.5
Potassium sulphate	K_2SO_4	174	87	17.4	31.1	24.8
Potassium chloride	KCl_4	74.5	74.5	14.9	26.6	21.2
Potassium phosphate	K_3PO_4	212.3	70.8	14.1	25.2	20.2
Ferrous sulphate	Fe_2SO_4	152	76	15.2	27.1	21.7
Ferric sulphate	$Fe_2(SO_4)_3$	400	66.6	13.3	23.8	19
Ferric chloride	$FeCl_3$	162.5	54.2	10.8	19.3	15.4
Aluminium sulphate	$Al_2(SO_4)_3$	342	57	11.4	20.3	16.3

Table 6.

pH	a CO_2/ac	a HCO_3^-/ac	a CO_3^{--}/ac
2.0	1.0000		
2.5	0.9999	0.0001	—
3.0	0.9996	0.0004	—
3.5	0.9986	0.0014	
4.0	0.9957	0.0043	—
4.5	0.9866	0.0134	
5.0	0.9587	0.0413	—
5.5	0.8800	0.1200	
6.0	0.6988	0.3012	
6.35	0.5000	0.5000	—
6.5	0.4232	0.5767	0.0001
7.0	0.1883	0.8113	0.0004
7.5	0.0683	0.9303	0.0014
8.0	0.0226	0.9728	0.0046
8.5	0.0072	0.9783	0.0145
9.0	0.0022	0.9530	0.0448
9.5	0.0006	0.8701	0.1293
10.0	0.0002	0.6801	0.3197
10.3	—	0.5000	0.5000
10.5	—	0.4022	0.5978
11.0	—	0.1754	0.8246
11.5	—	0.0630	0.9370
12.0	—	0.0208	0.9792
12.5	—	0.0067	0.9933
13.0	—	0.0021	0.9979
13.5	—	0.0005	0.9995

Table 7.

Traitement	Eau brute (1)	Floculation par polymères cationiques (2)	Clarification au sulfate d'aluminium et filtration (3)	Floculation cationique et traitement aux zeolites Na (4)	Floculation cationique et traitement aux zeolites acides (5)	Floculation cationique et déminéralisation avec résines échangeuses (6)	Polissage avec résine (7)	Traitement à la chaux seule à froid (8)	Traitement à la chaux seule à chaud (9)	Traitement chaux + CO_3Na_2 à chaud (10)	Traitement chaux à chaud + zeolite (11)
Impuretés en ppm de											
TA (en $CaCO_3$)	0	0	0	0	0	0	0	35	35	40	40
TAC (en $CaCO_3$)	235	235	208	235	0	4	3	70	70	80	80
Chlorures (en Cl)	11	14	14	14	14	0	0	14	14	14	14
Calcium (en $CaCO_3$)	144	144	144	0	3	0	0	35	30	10	0
Magnésium (en $CaCO_3$)	96	96	96	0	2	0	0	86	20	10	0
Dureté (en $CaCO_3$)	240	240	240	0,1	5	0	0	121	51	20	0,1
Phosphate (en PO^{---}_3)	1	1	1	1	1	0	0	0,5	0,1	0,1	0,1
Sodium (en Na^+)	5	5	5	115	5	0,4	0,01	5	21	21	30
Silice (en SiO_2)	6	6	6	6	6	0,02	0,005	6	2	1	1
Sulfates (en SO^{--}_4)	8	8	21	8	8	0	0	8	8	8	8
Nitrate (en NO^-_3)	5	5	5	5	5	0	0	5	5	5	5
Solides dissouts totaux	250	250	262	262	35	2	0,03	170	145	95	95
Solides en suspension	50	0,5	0,5	0,5	0,5	0,1	0,1	0,5	0,5	0,5	0,3
pH	7,8	7,8	6,8	7,8	7,8	9,0	7,2	10,1	10,1	10,1	10,1

SECOND PART

Cooling Water

Introduction

Thermodynamic characteristics of water and its availibilty in most industrial regions have made water an ideal fluid used for heat transfer in the industrial medium and power plants.

In the regions where water is rare, the cooling systems using air, of the type air-cooled battery, undergo a significant development.

Cooling water is widely used in most of the industries of developed countries, in particular, in thermal and nuclear power stations for vapour condensation; in iron and steel industry for cooling of facilities used for manufacturing of cast iron and steel, in chemical and petrochemical industries for the condensation of distillates, cooling of compressors, cooling of the products giving rise to exothermic reactions. These are the fields where the use of cooling water is the most considerable.

These uses represent very significant quantities of water:

— production of a tonne of steel uses 150 tonnes of water;
— production of a tonne of paper requires $25m^3$ of water;
— production of electrical energy requires $3m^3$ / MWh for the vapour and $60m^3$ /MWh for the cooling.

In total, these industrial consumptions worldwide are several times superior to the domestic consumptions, and this disproportion is all the greater since the country is more industrialized.

1

Conservation of the Raw Material: Water

The quantities of water involved in human activities are very significant. The demand of using water has only been increasing, whereas fresh water resources do not vary, being associated to the meteorological phenomenon which remains statistically more or less constant in most of the countries.

Single-purpose use of water is still widespread, although since some years there have been efforts to develop its re-use, for many reasons:

— scarcity of water: in an increasing number of places the demand is likely to be superior to the natural resources available;
— modification of the quality of water through the use, either the thermal modification (temperature rise leads to a modification of the aquatic flora and fauna), or chemical modification by introducing more or less harmful by-products of the human activities, which result in the destruction of the flora and fauna.

In economic terms, the use in straight-through passage of water is not always possible neither; water inlet into the facilities must have well-defined characteristics in order to be used in the industrial process, and this implies a treatment, which, though costly, can be made economical only when water is recirculated.

Finally, many countries have adopted regulations imposing on the industrialists to reduce their water withdrawal/intake from natural environment, thus to promote recycling and re-use of water.

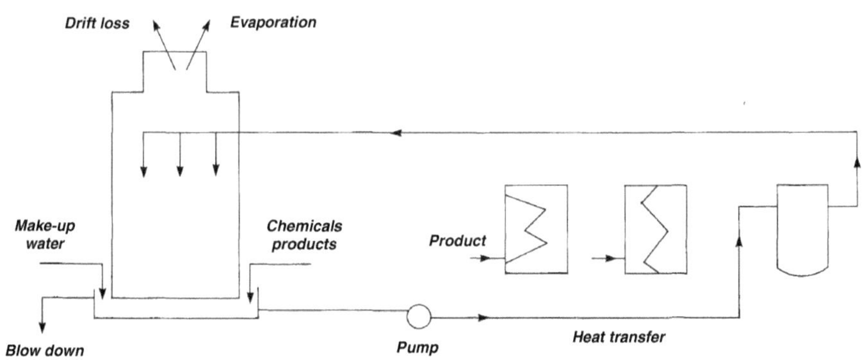

COOLING SYSTEM WITH TOWER

Energy consumption

Water	Energy	Heat transfer
Make-up	Electricity	Scale
Evaporation	Vapour	Clogging
Blow-down	Loss in product	Corrosion
Drift loss	Duration of manufacturing	Metals
Treatment		

2

Conservation of Energy

By this term we should understand not the thermodynamic principle of energy conservation, but energy savings, whatever the form, and the prevention of energy loss.

Conservation of energy may depend on the type of energy concerned, but at whatever point of an industrial production line, there is a need to ensure the energy management.

As soon as an equipment such as a heat exchanger, for example, does not operate in the conditions for which it had been calculated, there is an energy wastage, as either it does not fulfill its role perfectly, or it fulfills it completely, but with an energy expenditure greater than necessary.

Similarly: a product lost because of a defect in functioning of an equipment constitutes an energy loss: the energy which would be spent in order to synthesize the raw materials.

In the same manner, the reduction of the service life of an equipment due to corrosion implies an energy loss, since it would be necessary to invest again earlier to buy a new equipment. In all the cases, the energy management can directly be estimated in materials economy terms, with relevant monetary value.

Let us quickly examine different cooling systems using water, in order to identify there the points of energy consumption, thus places where energy savings may be realized.

Fundamentally, a cooling circuit is meant to extract heat and to dissipate it. A typical system with water recirculation using cooling tower, also called "open recirculating circuit", is illustrated in figure 1 (page 68).

Cooling water is pumped from the tank/reservoir towards the equipment, for example, an exchanger where it collects heat, then goes towards the cooling tower where it is cooled by evaporation, for returning then to the tank/reservoir and start the cycle again. Each part of the system has specific and precise functions to fulfill, so as to produce an efficient system of heat balance. We will discuss each of these parts separately.

Other common types of cooling water circuits are:

- Open circuits, also called once-through circuits, with straight-through flow, in which all the water is discharged into natural environment after use;
- Completely closed circuits where the evaporation is nil (figure below). Losses are accidental. There is no make-up. Cooling of the water in the circuit is made by an exchanger cooled by water or by air.

Basically, operation of the cooling systems is the same. Water enters in a heat exchanger where it collects heat. Then it goes out from it (as in the completely-open cooling system), to be replaced by "fresh" water, or it flows towards another heat exchanger where it delivers the heat that it had initially collected (as in the open recirculating circuit with tower). Water then returns to the tank, to repeat the cycle.

We will discuss at first the problems encountered during such kind of treatment, their effects and relevant solutions, rather than the types of cooling circuits.

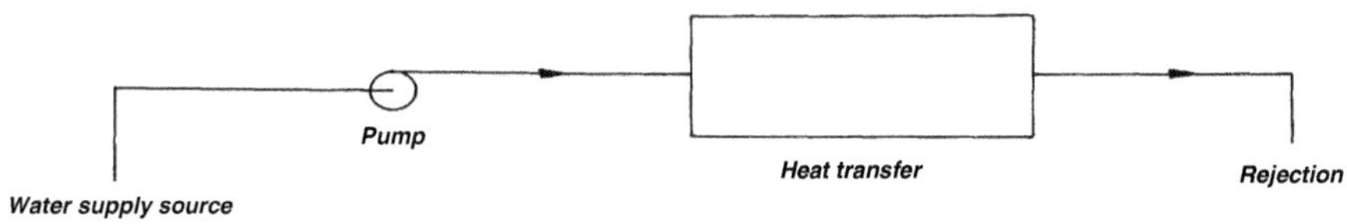

SYSTEM WITHOUT REGARDING

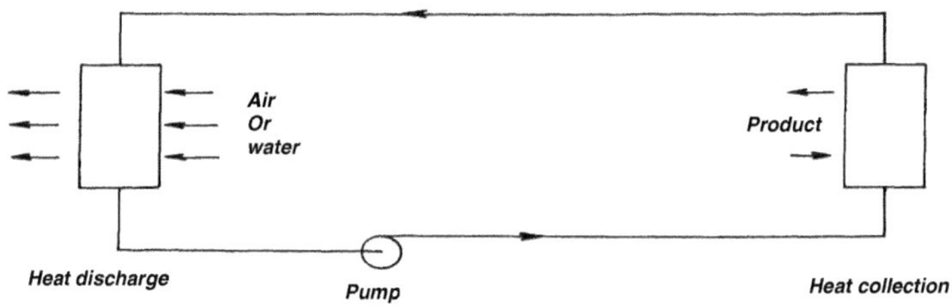

CLOSED SYSTEM WITH REGARDING

3

Conservation of Water and Treatment

Water is the element common to these circuits and therefore, logically, the most apparent and often most immediate source of energy savings, that could be introduced. This is first applied to the open recirculating systems with cooling tower, but may also be entirely true of the open circuits with single passage, as well as the completely closed circuits.

The conservation of water means the conservation of energy in the initial water treatment, pumping, treatment of water circulating in the wastewater treatment system. We will also examine what does it mean in a typical open recirculating cooling circuit with tower and what is about the amount of water needed. Let us consider the case where the recirculation throughput is of 10,000 m^3 per hour, with a difference of temperature (resulting from passage) through the cooling tower of 10 °C (change from 43 °C to 33 °C). By taking an average evaporation of 1 % for every 5 °C of temperature fall, we obtain the following table concerning the amounts of water needed for different concentration ratios.

Water needs of a cooling system with tower

(Recirculation: 10,000 m^3 per hour - $\Delta T = 10$ °C)

Cycles of concentration	Make-up water (m^3/hour)	Evaporation (m^3/hour)	Blow-down	
			m^3/hour	m^3 per day
1.2	1200	200	1000	24000
1.5	600	200	400	9600
2	400	200	200	4800
3	300	200	100	2400
4	270	200	70	1680
5	250	200	50	1200
7	230	200	30	720
10	220	200	20	480

This table shows well the savings that we can realize on the consumption of make-up water, and consequently, on the energy of pumping, as soon as we concentrate the recirculation water. We also see that beginning from a certain concentration ratio, the saving realized in terms of water is low and, as, in general, we would increase the problems of deposit on heat exchangers, by increasing the concentration ratio, it will be necessary to fix a limit to respect in order to reconcile these two parameters of operation.

The cost of water can be lower or higher depending on its origin. Municipal water is in general the most expensive. If we have a plentiful source of supply, which does not necessitate any treatment, then with only costs of pumping and distribution, we would have the most economical water. Purified water can be re-used.

Savings regarding the water expenses apply not only to the make-up water, but also to the blow-down water which may necessitate a treatment before its discharge into natural environment.

Moreover, there is a need to realize savings on the handling and maintenance of the equipment of supply and control, as the costs concerned can be substantial. Modern and reliable tools are now available in order to manage, analyse and control the cooling circuits: remote surveillance with alarm, automation of treatments, continuous automatic control of different parameters, management of the stocks of reagents.

There are other forms of savings, quite real concerning water, in particular, the re-use of the wastewater as make-up water. The industrial or municipal effluents, once treated, may be re-used under certain conditions. This is the case in certain number of public service enterprises, refineries, chemical and steelworks factories in Europe and in the United States. This is a promising field to explore, by reason of the water shortage.

► *Heat transfer savings*

Probably, one of the most significant losses of energy in the operation of a cooling circuit is the one caused by a loss in the exchange capacity. This is produced when there are deposits on the heat exchange surfaces. These deposits can be salts of calcium and magnesium, compounds of iron (corrosion) and aluminium, sludge and fouling issued from the process leakage or of microbiological origin (gelatinous or viscous) or corrosion products.

Typical deposits and their sources are listed in the following table:

Deposits of natural origin	
Issued from water	*Issued from air*
Fouling Sludge Natural organic matters Dissolved minerals Microbiological and macrobiological organisms	Gas Dust and ground dust Organic matters (vegetation) Microbiological organisms Insects, etc.

Deposits of artificial origin		
Issued from water Particles carried after passage of the water in the clarifier Pollutants Phosphates Detergents Sewer effluents	*Issued from air* Organic gas Hydrogen sulphide Sulfur dioxide (SO_2) Carbon dioxide (CO_2) Ammonia	*Issued from the circuit* Corrosion products Inhibitors of corrosion and their reactants Process pollutants

Deposits are source of energy loss owing to a reduction of the heat transfer capacity, an additional energy consumption of pumping and may, in extreme cases, necessitate the interruption of the equipment operation in order to allow its cleaning or replacement of the equipment, with the operation loss resulting from it.

Causes of deposits may be various:

- — carry-over of particles coming from the clarifier (post-flocculation);
- — excessive concentration of suspended solids;
- — defective pH control;
- — incompatibility of the water treatment products;
- — pollution/contamination caused by the process;
- — atmospheric pollution;
- — too low passage velocity;
- — excessive concentration of inhibitor;
- — insuffcient concentration of inhibitor;
- — variations in the make-up water quality;
- — bad functioning of the equipment.

▶ *Maintainance savings*

These savings are those realized on the expenses related to:

a) time of immobilization;
b) cleaning materials and labour;
c) reduction in the service life of the equipment;
d) and/or premature replacement of the equipment.

Time of immobilization is first of all the time lost for the manufacturing. It also includes the costs of unproductive equipment and labour. These costs are usually counted in the production costs. However, if they are higher than those estimated, the cost of the product increases, often in large proportions. Methods to calculate the real costs of the time of immobilization and operation losses are specific to each industry, even each plant.

Costs of cleaning materials and labour vary depending on the type of cleaning carried out and the type of the deposit formed, as well as on whether the equipment should be dismantled for cleaning or can be cleaned without dismantling it.

Premature replacement of the equipment may become necessary because of an excessive accumulation of deposits or an excessive corrosion.

▶ *New systems*

We have first discussed the existing systems. Setting up of a new system may be an occasion as soon as its design proves orginal and capable to realize considerable savings. For example, we can consider the reduction of clogging factors in the specifications of heat exchangers. This may be realized with the results of substantial reduction of the cost of heat exchangers. This result may be obtained by operating with cleaner surfaces of exhange, thanks to the use of specific chemical products or mechanical equipment, or with both of them. Appropriate choice of materials during the realization of the exchange tubes may also bring substantial savings. However, attention must be paid to the manner in which we associate different metals in the heat exchangers or in the circuit, as it may reduce the service life of the equipment, due to the galvanic corrosion or may increase the rate of deposition, thus making impossible any energy savings, which had been planned or foreseen.

To sum up, it is possible to realize considerable energy savings in the cooling water circuits. Each circuit should be examined carefully as a whole, with all the sections of the plant or workshops concerned, so as to obtain the savings due to water, chemicals, operation of the system or even due to the system design.

It is necessary to focus on both the obvious actions and those which are apparently and relatively insignificant or "impossible to realize". Each cooling circuit may bring energy savings which are realizable often for a minimal cost and without real expenses.

4

History of the Cooling Water Treatment

Let us to remind that the oldest use (that we know) of a corrosion inhibitor dates back to the year 1905, when it was found that arsenic was capable to prevent the dissolving of iron in the hydrochloric acid. The first mention concerning an inhibitor for water dates from 1922: some ppm of sodium silicate was added to water of domestic use in order to avoid corrosion of the piping made of lead. The first mention concerning cooling water treatment dates back to 1924, when 200 ppm of sodium dichromate and 50 ppm of sodium hydroxide were used in a refrigeration plant. Thus, as we can see it, the history of the cooling water treatment covers less than 100 years. Most of the significant developments in this field occured in the course of the professional career of many chemical engineers which are still in office.

At the beginning, the enemy was scale. It was fought by the use of external coolers, equipment of a huge size, and by frequent cleaning which implemented a mechanical removal of the scale. The only treatment which had been used almost commonly was the occasional addition of a certain amount of copper sulphate, meant to remove algae sufficiently, in order that the tower does not collapse due to the increase in weight caused by the proliferations of micro-organisms.

Difficulty of the problem consisted in the fact that almost nothing was known as regards the mechanism of the scale formation. The only thing that was then known is that most of calcium salts had an inverse curve of temperature-solubility, making preferable the use of the water of low hardness. It was also known that the scale is formed more rapidly at the higher pH and, on occasion, acid had been added to water. However, no tower operated at a pH inferior to 7.5. It was known that certain organic compounds deform the structure of scale crystals, thus making it less adhesive. Tannin and lignin had thus been used sometimes.

The first breakthrough was the development of the solubility index by Langelier in 1936, and when the parameters influencing the scale formation were determined: calcium hardness, total alkalinity, pH, concentration of dissolved salts and skin temperature.

However, every solution brought to the scale problem had only been revealing new problems. In reality, what the Langelier index meant is that: "For a given set of conditions, all the natural waters either form scale or corrosive".

Two fronts were now open, in order to solve the problem. The first consisted in adjusting the pH of water, usually by adding acid, so that a fine protecting layer of scale settles: it is a matter of the method named "controlled scale deposit". In theory, it is a good method. In practice, it is impossible to be applied. When the hot part of the heat exchanger is protected, the cold part is attacked.

When, on the contrary, the cold part is protected, the hot part is covered by a (very) considerable layer of scale. However, despite its limitations, this method was useful, as it allowed increasing up to 50 % the running time of the heat exchangers (possibly, nine months between two successive cleanings).

The second method consisted in treating water so as to make it deliberately corrosive, and then to add a corrosion inhibitor. The corrosion inhibitors then known were inefficient, unless added in quantities (which are) prohibitive from the economic point of view. Chromates restricted the uniform corrosion, but provoked an attack by pitting. Polyphosphates restricted the uniform corrosion, but they turned into orthophosphates, which provoked a new scale deposit.

The second breakthrough took place during the Second World War. The English had discovered that if two corrosion inhibitors were added to water, the anti-corrosive action of the combination of the two substances were well superior to the sum of their individual actions. This is what is called the "synergetic effect". The particular combination that the English used was that of the polyphosphate and the nitrite. At the beginning, such combinations were strictly empirical. Much later, it was discovered that a cathodic inhibitor and an anodic inhibitor should have been used. This discovery was considered as being of such an importance that it was ranked as top secret. Its use was refused not only to the ennemies, but also to the Allies.

Shortly after the war, the secret was divulged. The first product which appeared on the Amercian market was a polyphosphate-ferrocyanide formula. It was followed soon after by a polyphosphate-chromate formula. A bit later, zinc and certain organic compounds were introduced to the market, in their turn. Engineer working on the corrosion problems had at his disposal many combinations: chromate-zinc, phosphate-zinc, chromate-phosphate-zinc, zinc-organic compound, chromate-phosphate-zinc-organic derivative, etc.

Then, new problems had emerged, again. Cooling water circulated at the pH of 6.0 to 6.5. These conditions of pH, added to the high hot humidity and the absence of light, were favourable to the growth of fungi in the cooling towers, which were made of wood. Soon after, two destructive species made their appearance in the cooling towers. It was a matter of the "white rot", which attacked the lignin, leaving behind it the cellulose of weak structure. This attack often occured within the framework pieces supporting the construction, leaving for the posterity a long history of collapsed towers. The "brown rot" attacked the cellulose, which provoked the failure of the lignin, and then, the subsidence of the walls of relevant devices.

Sometimes, the two rots occured in the same time, and the construction wood of the cooling towers just simply disappeared. No treatment of water was efficient, since the attack occured in the non-submerged parts of the tower.

This problem was solved in two manners: firstly, materials other than wood started being used for the construction of the cooling towers: asbestos cement, stratified plastic, ceramics, etc. Secondly, woods were treated before their use for the construction of the cooling towers. Wood was impregnated with a copper salt, then, with an arsenic salt. This was called the treatment "with double diffusion". The copper arsenide precipitated in the wood block provided a good protection. Otherwise, wood was impregnated with creosote. Both methods have had their own followers and opponents. However, both of them appeared satisfactory. Treatments were carried out at the site of the existing towers, where the wood had never been treated. This prolonged their service life. However, finally, it was necessary to replace the wood.

The second problem was that of the sulphate-reducing bacteria. These bacteria, again, can exist only in the water of low pH. They are anaerobics.

However, they are capable of coating (encapsulation) and they can subsist in the heat exchangers, even in the presence of oxygen-saturated water. They thrive on sulphates (introduced into the system in the form of the sulphuric acid), by liberating hydrogen sulphide. Two unfortunate consequences result from this: the biological mass, hung to the metallic surface reduces the heat exchange. The hydrogen sulphide produced is found in direct contact with the metal, which provokes its corrosion, while the biological mass isolates from the metal the corrosion inhibitor. Very often, other organisms producing organic sludge are present as well.

The oldest method to fight against these microorganisms, as well as the algae, consisted in administering periodically some biocides to the system. Usually, two different biocides had been used alternately once a month. The reason of so doing was to prevent the microorganisms from getting used to a single biocide. Copper citrate and chlorophenates were the first favourite ones. Later, tributyl-tin oxide replaced the copper citrate.

This method worked rather well, but the biocide shock treatment has the disadvantage of leaving the system without treatment during a quite long period. Moreover, the continuous addition of these biocides was not economical.

Another solution consisted in sterilizing completely the system by a continuous treatment with chlorine. However, the concentration required (about 1 ppm of residual chlorine) attacked the wood in the cooling tower, especially in the sunlight. This attack shortened by half the service life of the lattice, and wood fibres present in the water in circulation had considerably aggravated the problem of clogging (fouling).

Therefore, a compromise was necessary. A potential compromise was the shock treatment with chlorine (a residue of 1 ppm was made up and water was recirculated during about 4 hours, three times a week). Another method consisted in making circulate, in continuous manner, a small amount of chlorine, about 0.1 ppm of residual chlorine. This treatment with chlorine was reinforced by a periodic addition of quaternary ammonium compound salts.

The quaternary ammonium compound salts have a weak bactericidal activity, but their detergent action is capable of removing the colonies of sulphate-reducing bacteria, so that oxygen and/or chlorine could complete the removal. Occasional addition of methylene-bis-thiocyanate, meant to control the algae, was useful as well.

The third problem was slower to appear. Water chemists had rather been satisfied with themselves due to the fact that the limiting factor for the cleaning of heat exchangers was no more the scale, the water side, but clogging or fouling, concerning the fluid of the process. However, in the 1950s, there was a progress thanks to the use of inhibitors and detergents, concerning the fluid of the process, thus reversing the trend, and the water side became again the critical one.

This time, the enemy was clogging (fouling), which consists in the deposition of a suspended solid, whereas the scale is the deposit of salts which are usually found in the solution.

Colloidal substances may penetrate in the water circulating in the circuit in different ways. They are found in make-up water, especially when it is a matter of the untreated surface water. They may come from the air during its contact with water in the tower. They may also be caused by the deterioration of the materials used for the construction of the tower (packing) or the algae present in the tank. They may as well have as origin the reaction products connected with the treatment, mainly in the form of iron, phosphate and calcium sulphate.

These colloidal particles have a tendency to take a negative charge in acid water, whereas the cooling surfaces made of steel have a tendency to be charged positively, which leads to their deposition and thus favours the fouling. Higher temperatures favour this process which is very simplified.

The colloidal matters present in feed water can be removed by coagulation and filtration. Their abundant accumulation in water in circulation can be controlled by means of the sidestream filtration. The control of the development of microorganisms has already been discussed above. The problem of fouling or clogging due to the phosphates can be resolved by replacing them by the corrosion inhibitors belonging to other chemical families: zinc, molybdenum, etc. The use of the anionic dispersants improves the results as well.

The last breakthrough, and may be the most important in the field of cooling water treatment, is indeed a double breakthrough.

Around 1968, a series of organic compounds of phosphorus, the phosphonates, were developed. These elements have the property of conserving calcium salts in the solution, even at a pH of 9.0 without addition of any acid. Some of them have the anti-corrosive properties as well. They allow the use of corrosion inhibitors of lower concentrations.

This operation, carried out at the higher pH, also reduces the possibility of the presence of sulphate-reducing bacteria in the circuit.

Almost simultaneously, a series of anionic polymers had been introduced into the market. These are the polyacrylates, whose use reduces considerably the tendency to clogging or fouling of the water circulating in the cooling towers.

As usual, the solving of an old problem reveals a new one. And the new problems necessitate new solutions, to the point that the treatment of water circulating in the cooling towers will remain an art.

Currently, there are five main axis of development of the cooling water treatment. These are the following:

— reduction of the water consumption;
— prolongation of the service life of the equipment, by stabilizing the performance of thermal exchanges at the nominal values considered by the plant designers;
— limitation of the impact on environment caused by the waste/discharge;
— safety of the personnel which is on contact with the products used for the treatment;
— control of the legionella risk for the population exposed to aerosol generated by the evaporation of water.

5

Cooling Circuits

5.1 Composition of the Circuits

Its great availibility and its remarkable thermodynamic properties make water an ideal fluid for realizing heat transfer necessary to the good functioning of the industrial processes.

The main elements constituting an open recirculating circuit are:

— devices to be cooled: condensers, heat exchangers, coolers of various fluids, heat engines, compressors and different units, from which it is necessary to evacuate calories (blast furnaces, furnaces, chemical reactors, etc.);
— one or several cooling towers (atmospheric cooling towers), whose role is to remove the heat transferred to water, in the devices cooled, by direct contact with air;
— one or several pumps which take cold water from the tanks near the cooler base, to send it towards the equipment to be cooled;
— connecting pipes and related devices such as the one for sidestream filtration.

5.2 Heat Transfer

Heat transfer simply consists in transferring heat, from a unit to be cooled or from a process, towards a receiver which is the cooling water.

The cooling water is usually not in direct contact with the process that needs cooling. The heat transfer is made through a wall, most often metallic, which should be a good heat conductor.

These heat transfer surfaces constitute the main part of heat exchangers.

There are different types of heat exchangers, depending on the configuration of the exchange surfaces and the nature of the fluid to be cooled. In the case where we cool and condense the water vapour, the device is called condenser; if the receiver is a liquid that we should vaporize, the exchanger is called evaporator.

The types of heat exchangers of the most widespread use in industries are the following:

— Tube heat exchangers: a bundle of tubes in which the process usually passes is situated inside a shell: cold water circulates through the space situated between the (external side of the) tubes and the (internal side of the) shell. There are many variants of this type of exchanger: simple passage or double passage, etc. The number of tubes and the surface developed may be quite large: several thousands of tubes for a condenser of a thermal power station.

The tube bundle can be made of plain carbon steel, stainless steel, copper alloys and other alloys, which are good heat conductors. The choice depends on the nature of the fluid to be cooled and the economic conditions. The shell is most often made of the plain carbon steel. The following figure shows the cooling realized in an exchanger with a single backflow/countercurrent passage. Here, only one tube is represented.

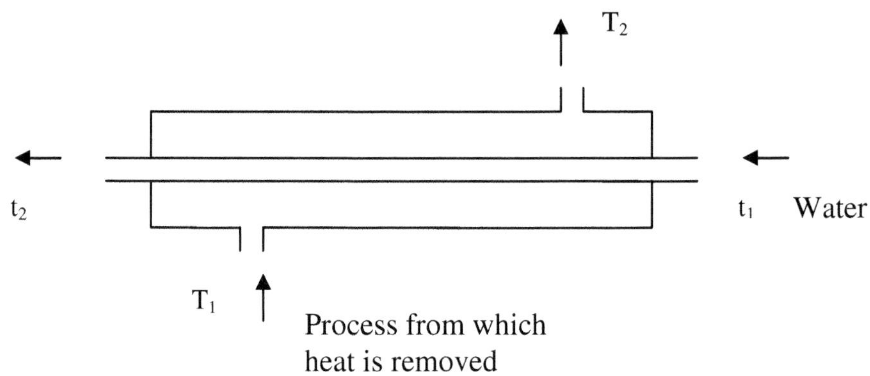

SIMPLE TUBE HEAT EXCHANGER

— Plate heat exchanger: here, the exchange wall is not constituted by tubes, but by plates/sheets. This type of exchanger is usually made of stainless steel.
— Reactor with double shell used in the chemical industry is another simple type of heat exchanger, where cold water circulates in the space separating the two walls of the reactor, which allows removing heat generated by the chemical reaction.

The heat transfer through the exchange surface is made by conduction, from the process to the metal, and then, from the metal to the cooling water. The heated water can then be cooled by convection, by mixture with the cold water.

The parameters important for realizing a heat transfer from a cold source to a receiver by conduction are the following:

— heat transfer characteristics of the exchange surface (heat conductivity);
— thickness of the wall of heat transfer;
— exchange surface;
— temperature difference between the process and the cold water;
— cleanliness of the exchange surface, concerning both the process and the water. Deposits constitute a barrier for heat transfer by conduction.

The following table presents heat conductivity values of different metals and alloys used in the manufacturing of heat exchangers, as well as those of the chemical and biological compounds playing the role of insulating materials.

Material	heat conductivity (W/m·K)
carbon steel	45
stainless steel	15
pure copper	384
Aluminium	200
brass with 30% of zinc	99
calcium carbonate	2.6
calcium sulphate	2.3
ferric oxide	2.9
biofilm	0.6

Reduction of the heat conductivity due to the deposits causes losses in production and necessitates the increase of the cooling water throughput.

We can monitor the heat transfer quality of an exchanger using its transmission coefficient given by the following formula, by referring to the figure of preceding page.

$$K = q / S \times \Delta T_M,$$

where: $\Delta T_M = (T_2\text{-}t_1) - (T_1\text{-}t_2) / \log_e [(T_2\text{-}t_1) / (T_1\text{-}t_2)]$
and $q = Q\ (t_2\text{-}t_1)$

K = transmission coefficient (kcal/m^2·h·°C)
q = thermal load (kcal·h^{-1})
ΔT_M = mean temperature difference between two sides of the exchanger wall
S = exchange surface area (m^2)
t_1 = temperature of the inlet cooling water (°C)
t_2 = temperature of the outlet cooling water (°C)
T_1 = temperature of the process before cooling (°C)
T_2 = temperature of the process after cooling (°C)
Q = cooling water throughput (m^3/h)

We have already said that 3 types of cooling circuits may be encountered.

5.3 Open or Once-through Circuits

These are circuits without recirculation of water or systems with a single passage (once-through system) or streamline system. We will not describe them, since they do not have specific equipment, as the water heated by the industrial process is discharged into natural environment. At most, there is additional equipment meant to filter the water or to add there the treatment additives from time to time. This system is widespread when the quantity of water available is large enough. As we have already said it, this type of system has been used less and less, due to the fact that it requires significant extraction of water, as well as the increase in temperature of the receiving body of water. A simplified diagram of a once-through circuit is represented in the following figure.

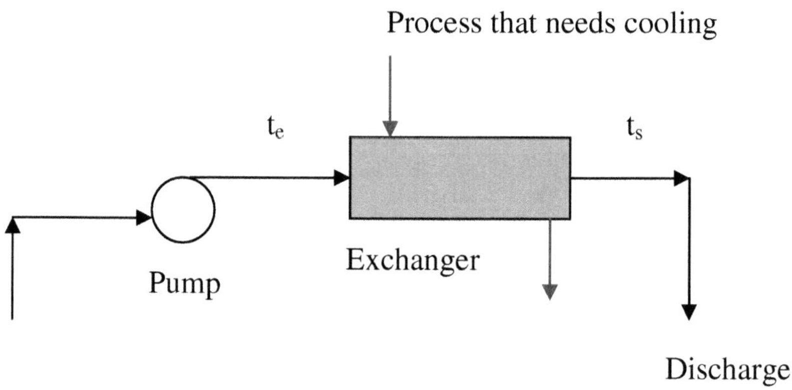

Open circuit

The circuit is characterized by the cooling water throughput Q and the difference between water temperatures at the inlet and the outlet of the equipment cooled:

$$\Delta T\,(°X) = t_s - t_e$$

One of the main characteristics of this type of circuit is relatively large quantity of water which is usually necessary for the cooling. The most frequent problems encountered in such kind of circuits are: fouling and scale deposits.

5.4 Open Recirculating Circuits

In such circuits, water is recycled after cooling, through passage in a cooling tower, in which it is cooled, in its turn, by contact with the air. The use of this type of circuit is widespread in industries, thus a whole chapter is dedicated to it. A simplified diagram of the open recirculating circuit is represented below.

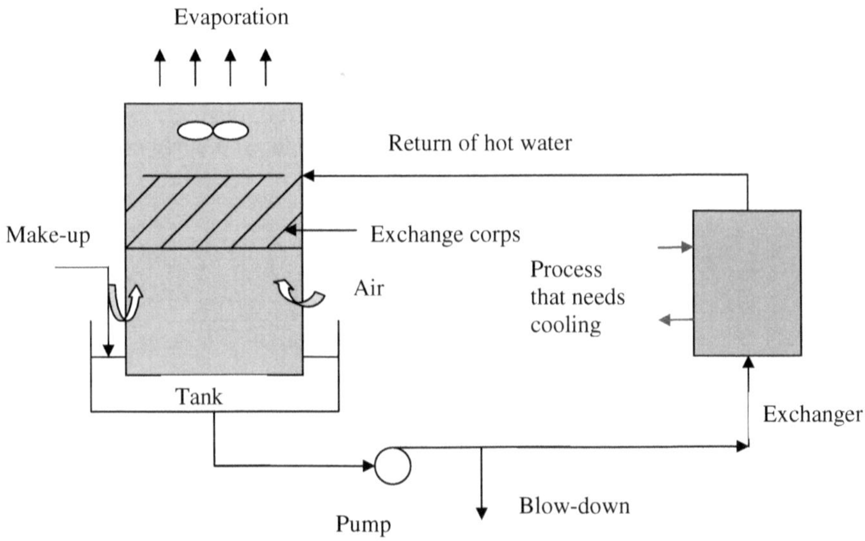

Simplified diagram of an open recirculating circuit

There are different types of cooling towers, usually classified by the air flow generation methods, also called "draught". Draught can be:

- **natural:** in the natural draught cooling tower, also called "hyperboloid" because of its form, the circulation of air is produced countercurrent in relation to the water to be cooled. Towers of this type, usually of very large sizes, are found in the thermal and nuclear power plants.

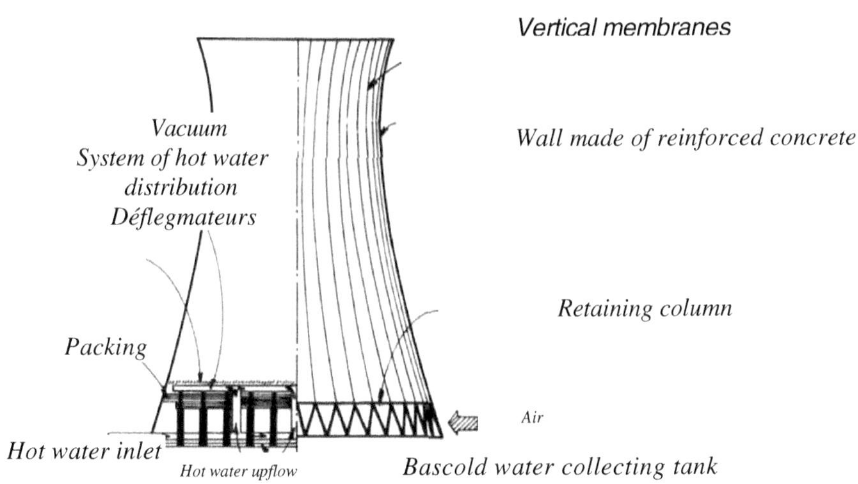

Countercurrent hyperbolic cooling tower

- **forced:** in the forced draught cooling tower, the air circulates in the tower under the effect of a mechanical ventilation provoked by a lateral fan which forces the air into the tower.
- **induced:** in the towers of this type, the air is pulled through the tower by a fan located at the top (discharge) of the tower. The induced draught cooling tower is widespread.

Induced draught cooler with three cells (doc. Scam - Alsthom)

In the forced or induced draught towers, the air circulates usually countercurrent in relation to the water to be cooled. Other variants can also be found: cross-current, etc. Manufacturers of the towers propose modular coolers, placed in series, which allow adapting cooling needs to the conditions of operation.

5.5 Closed Circuits

The closed circuits are in fact constituted by two thermally connected systems; one of them, which is completely closed (except leakages) serves for making circulate water and remove heat from the process through a heat exchanger, and the other, also called "secondary circuit", serves for cooling the water heated. The secondary circuit is quite often an open recirculating circuit or a system of air-cooled batteries (air cooling). The most famous example is the cooling of heat engines with closed circulation of water, where the cooling of water is made through an air-cooled heat exchanger (radiator in a car). The advantage of this process is that, theoretically, it does not consume any water. It is also quite easy to maintain the homogeneous composition of the water at all the points of the circuit. There are two disadvantages:

— the cooling device is quite volumunous;
— the "low" temperature obtained is relatively high.

The figure below refers to a simplified diagram of this type of circuit.

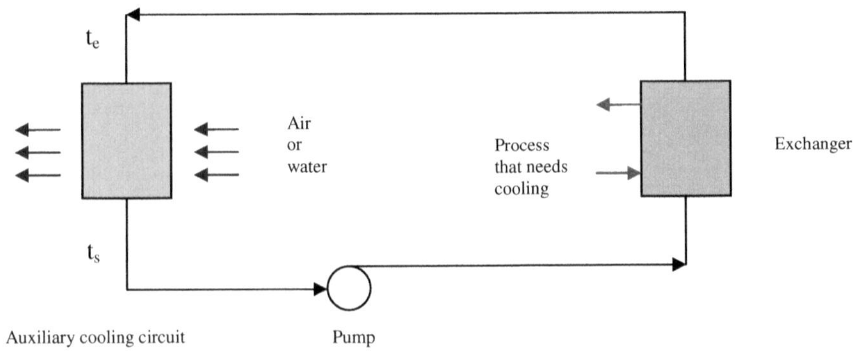

Diagram of the closed circuit

As a circuit called "closed" is not completely closed in practice, it is necessary to apply the water make-up, preferably, of softened or demineralized water. These facilities are usually made of different metals, and this presents a high risk of galvanic corrosion.

Therefore, anti-corrosive measures are taken. Following three treatments are frequently used:

— alkaline and oxygen-reducing treatment;
— alkaline treatment with a nitrites- molybdate mixture of high concentration, to which a dispersant is added.
— treatment with an organic corrosion inhibitor in alkaline environment.

As the contact with the air or any other element likely to cause a biological contamination is minimal, the risk of bacterial proliferation is practically nonexistent in the closed circuits.

In a sufficiently tight closed circuit, operating with softened or demineralized water, the formation of scale deposits on exchange surfaces is quite limited, even impossible.

When the make-up is significant, the risk of deposit formation appears again: scale, corrosion products due to the oxygen dissolved and fouling by microorganisms, and the latter phenomenon may be amplified by process leakages.

6

Fundamental Principles of Cooling Towers

6.1 Operation Principles

We have seen that an atmospheric cooling tower is a heat exchanger, in which the water to be cooled flows in contact with the ambient air, making it trickle on piles of superimposed screens or plates (dispersion area). It is the packing that constitutes the heat exchanger. It is closed in a tower ensuring the best possible air circulation.

The function of this cooling tower is only to cool hot water. The water cooled in this way is made circulate until the unit in operation, where its role is to remove the heat produced in this unit. Thus, the previously cooled water becomes hot again and is sent back to the cooling tower in order to be cooled again and recycled.

The water is cooled by means of a tight contact with the air through two processes:

— conduction: the heat coming with the hot water is transferred to the cooler air by contact. This represents about 15 to 25 % of the total heat exchange;
— evaporation: the weight of the water vapour per unit of time and surface is proportional to the difference between the partial pressure of the water vapour in the atmospheric air and the saturated (water) vapour pressure of this air at the temperature t.

This difference of partial pressure itself is proportional to the difference between the weight of vapour per kg of the dry air in contact with the water, and the weight of vapour per kg of the saturated air at the same temperature.

In practice, the quantity of the water evaporated per hour is situated between 1 and 3 % of the quantity of the water circulated, for a difference of temperature of 10 °C, and this quantity varies depending on the hygrometric state of the air, the inlet water temperature, was well as the the cooler design.

► *Losses caused by mechnical drive and drift/windage loss*

The tight contact between the air flow and the water flow gives rise to a drif/windage loss, which is controlled by a drift eliminator or droplet; nevertheless, this loss exists and can be quite variable, from 0.5 to 0.0005 % of the recirculated water flow. An efficient separator and a regular maintenance of the tower allow reducing the droplet emission rate. This loss does not contribute to the cooling.

The drift/windage loss has a highly considerable role in the spread of bacteria of the type *Legionella*.

With a view to maintain a relatively constant volume of water circulating through the system, the losses by evaporation should be compensated by a fresh water input.

1 Metallic framework
2 Moto-blower unit
3 Droplet separator
4 Water distribution
5 Exchange body: « packing»

Structure of atmospheric cooling tower (Doc. Scam-Alsthom)

► *Concentration of dissolved matters in water*

If the water used was absolutely pure water, that is to say, containing no dissolved solid, it would suffice only to replace the water evaporated and the problems would have been reduced to the minimum.

However, entirely pure water does not exist in the nature and it is too expensive to produce it. Natural waters contain dissolved or suspended salts, dissolved or suspended organic matters, as well as dissolved gases, all of which cause problems.

Let us take, as example, a hypothetical cooling system having a water volume of 1,000 m^3, and a recirculation rate of 2,000 m^3 per hour. Its evaporation rate is of 40 m^3 per hour (about 10 °C of temperature drop through the cooling tower).

If we suppose that the make-up water contains 400 ppm of dissolved solid matters, we have to start with 1,000 m^3 containing 400 kg of solid matters.

During the first twelve hours of operation, about 480 m^3 of water is evaporated, leaving the dissolved solid matters in the cooling system. And during this time, 480 m^3 of make-up water was added, which contains an additional 192 kg of dissolved solid matters. The amount of solid matters in the system now reaches 592 kg. The concentration of dissolved salts has increased by 50%, and after 24 hours of the similar operation, there will be about 800 ppm of dissolved salts; after 72 hours, there would be 1600 ppm of dissolved salts.

This increase in concentration can not be maintained, as the amount of (dissolved) salts will inevitably exceed the amount related to their solubility limit, thus causing deposits in the circuit, as well as the risks of corrosion under these deposits.

With the aim of avoiding such a concentration, a part of the water is continually removed from the circuit, in the form of blowdown. A small quantity, less than 0.1% of the recirculation water is carried over into the air. In this discussion, we will neglect the windage loss, although it is an important factor in the spread of microorganisms, particularly, of the *Legionella* bacteria, as we have already mentioned it above.

6.2 Elements for Calculation of the Cooling Towers with Evaporation

6.2.1 Notion of Concentration Cycle

We have seen in the previous example that by starting with water containing 400 ppm of dissolved solids, we will reach, after 72 hours of operation with constant volume, to the concentration of 1600 ppm of dissolved solids. The total concentration of dissolved solids (TS) has been multiplied by 4, that is to say, 4 concentration cycles.

6.2.2 Expression of Blowdown in Percentage

In order to maintain the concentration of dissolved solids of 1600 ppm, it is necessary to replace a part of the water with 1600 ppm of concentration by the initial water with 400 ppm of concentration:

The quantity to be replaced is: $P = 400 \times 100 / 1600 = 25\ \%$
P = blowdown percentage

6.2.3 Definition of the Terms Used for the Calculation

Q: Recirculation rate of the water in m^3/h. It is estimated depending on the data written down on the plate of recirculation pump.

ΔT: water temperature difference between the inlet and the outlet of the cooler, in °C

W: power of the cooler, in kcal/h (1 kW = 860 kcal / h)

E: evaporation rate in m^3/h. It depends on the quantity of water to be cooled, (Q) and the temperature difference, (ΔT). In empirical manner, we evaporate 1% of the recirculation rate (Q) for each drop of 5°C of temperature through the cooling tower.

E_v: windage loss in m^3/h

P: blowdown rate in m^3/h, adjusted in order to limit the concentration of mineral salts in the recirculation water, the concentration due to the evaporation of a part of the water recirculated.

A: make-up flow rate in m^3/h (compensates losses due to the blowdown and the evaporation)

V: volume of the circuit: in general, it is estimated by taking the volume of the tank, plus 20 to 50 %.

τ: half-exposure time = 0.7 x (V / P). It is the time in hour for diluting to 50 %, a substance introduced at a given moment.

R_C: concentration ratio between the salinity of the water in the circuit and the salinity of the make-up water.

In general, this ratio is determined by the dosage/measuring of the chlorides in the make-up water and the circuit water, and the verification of the stability of the calcium bicarbonate by the dosage/measuring and comparison of the ratios with the TH and the MA (methyl orange alkalinity).

As the Anglo-Saxon literature on the subject is quite abundant, we give the correspondences of some terms between the two languages.

FRANÇAIS		ANGLAIS	
Evaporation	E	Evaporation	E
Cycle de concentration	R_C	Cycle of concentration	C
Eau d'appoint	A	Make up water	M
Purge	P	Blowdown	B
Entraînement vésiculaire	E_V	Windage loss	W
Solides dissous dans l'eau d'appoint	$(SD)_A$	Total solids in make up water	$(TS)_M$
Solides dissous dans l'eau en circulation	$(SD)_C$	Total solids in recirculation water	$(TS)_R$
Volume utile du système	V	Holding capacity	HC
Indice de durée ou durée de vie	ID	Holding time index	HT
Fuites	F	Leaks	L
Débit de recirculation	Q	Recirculation rate	R

6.2.4 Number of Concentration Cycles

This number is not chosen just arbitrarily, but it is defined by the initial quantity of dissolved solids $(DS)_m$ and by the maximum tolerable concentration of dissolved solids in the circulating water $(DS)_c$.

$$C = \frac{(DS)_c}{(DS)_m} \tag{1}$$

It may be defined for each impurity in the water, which is likely to cause a problem. The number obtained for the impurity most rapidly causing a problem, thus the lowest, is that used in the calculations.

For the following calculations, we will assume a system in continuous function, that is to say, the value of $(DS)_c$ (during operation) is reached.

For any period of time considered, we have:

$$M = B + E + W \tag{2}$$

As we are in continuous operation, entering DS equals DS leaving.

$$M\ (DS)_m = (B + W)\ (DS)_c$$

Equation (1) gives:

$$(DS)_c = C \text{ x } (DS)_m$$

We obtain:

$$M \text{ x } (DS)_M = (B + W)\ C \text{ x } (DS)_M$$

$$M = (B + W)\ C$$

Thus:

$$C = M / (B + W) = (B + E + W) / (B + W) = E / (B + W) + 1,$$

hence:

$$B + W = E / (C - 1) \tag{3}$$

In practice, the windage loss is very small, and we can neglect it (in a well known installation).

We have now:

$$B = E / (C - 1) \tag{4}$$

$$M = B\,C \tag{5}$$

and we obtain:

$$M = E\,C / (C - 1) \tag{6}$$

6.2.5 Consequences

For a given installation, a higher tolerated concentration means less blowdown to be done, thus less make-up water to be supplied.

As an example, we have established the following table for an evaporation flow of 100 m^3/ h.

C	B	M
x 1.5	195 m^3/h	300 m^3/h
x 2	95 m^3/h	200 m^3/h
X 3	45 mVh	150 m^3/h
x 4	28 m^3/h	133 m^3/h
x 5	20 m^3/h	125 m^3/h
x 6	15 m^3/h	120 m^3/h
x 7	12 m^3/h	117 m^3/h

The following graph indicates the make-up water necessary, expressed in % of the recirculation rate, depending on the number of concentration cycles obtained or calculated for the cooling circuit obtained.

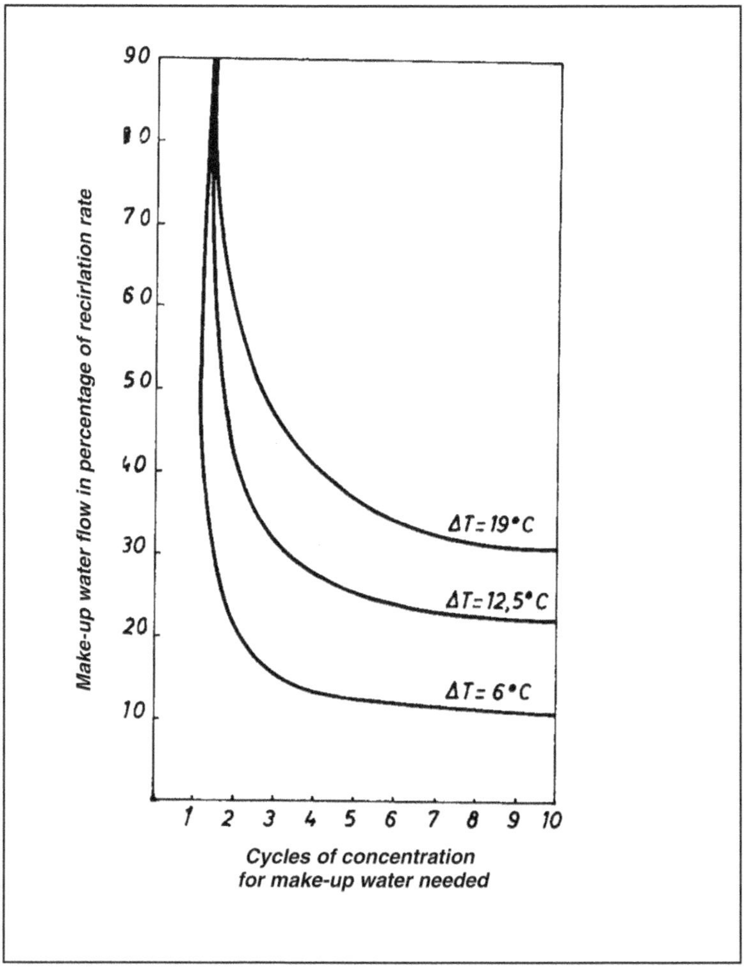

6.2.6 Calculation of the Concentration Rate

We have previously seen that all the parameters of a water circuit such as the make-up water, the blowdown, the evaporation and the drift loss are the rates/flows connected between them by the equation (2):

$$M = B + E + W \tag{2}$$

In practice, the drift/windage loss is very small and can be neglected in a well known facility/plant. The equation (2) then becomes:

$$M = B + E$$

Let us now see the change in concentration of a given ion in the circuit, for example, that of the calcium ion which is also found present in the make-up water. Its concentration in the circuit may increase or decrease depending on the blowdown rate.

If we carry out the calcium ion weight balance of the circuit, we would find that the modification of the concentration expressed in weight per hour of the calcium ion is:

$$\Delta M = V\,(Ca)_C - B\,(Ca)_C + M\,(Ca)_A \tag{7}$$

V: total volume of the circuit in m^3
$(Ca)_C$: concentration of Ca^{++} in the circuit, kg/m^3
$(Ca)_A$: concentration of Ca^{++} in the make-up water, kg/m^3
B and M: quantities of blowdown and make-up in m^3, removed and added in 1 hour.

Although the calcium ion concentration increases when the evaporation increases, the change in concentration can be expressed dependiong on the make-up and the blowdown, without considering the evaporation: Mac Coy (*), after having resolved the differential equation of the change in calcium concentration, proposes several formulas, among which the most frequently used one is so-called the equation of concentration (8).

$$(C)_t = M\,(C)_A\,/B + [\,(C)_0 - M\,(C)_A\,/B]\; e^{-P(t-t0)/V} \tag{8}$$

$(C)_0$ is the initial concentration of Calcium, at the moment t_0;
$(C)_t$ is the concentration of Calcium at the moment t.

This equation allows, in particular:

- determining the concentration after a given period of time;
- determining the time necessary for obtaining a given concentration.

In the same manner, Mac Coy (*) has also introduced the concentration decrease rate of the chemicals added, for example, that of zinc salts. It is important to know it, as it allows compensating, continuously or at given intervals, the loss of the substance and thus keeping its concentration at the desired value.

It is to be remarked that this rate is independent from the evaporation and from the make-up water flow, as the loss is caused only by the blowdown.

Mc Coy propose several formulas:

$$\log_e ((C)/(C)_o) = - B (t - to)/VU \qquad (9)$$

$$(C) = (C)_o \, e^{-p(t-t0)/V} \qquad (10)$$

These formulas allow anticipating the change in the concentration of a given additive depending on time.

(*) Mac Coy: book mentioned in the bibliography.

6.3 Necessity and Purpose of Cooling Water Treatment

In addition to the fact that it cools the operation unit, the recirculation water can also generate other actions/effects, most often undesirable: it can produce the scale, corrode the heat exchangers, internal coatings, pumps, valves, etc., attack the construction materials of the cooling tower, create an environment favourable to the growth of microorganisms, generate the aerosol containing bacteria of the type Legionella and contribute to the pollution problems.

In order to avoid or minimize these problems, the water should be treated.

1° In order to avoid the scale deposit on the exchange surfaces:

A. By eliminating, thanks to an external treatment, the elements producing scale:

- carbonate removal through the process of lime softening;
- softening through the ion exchange resins.

B. By keeping in solution, through an internal treatment, the elements producing scale:

- by reducing pH with an acid;
- by forming complexes with a sequestering agent;
- by anti-sprouting agents which blocks the crystalline growth during a period sufficient for avoiding the formation of mineral deposits.

2° In order to avoid the attack of metals in contact with the cooling water:

- by means of corrosion inhibitors adapted to the metals present;
- by means of a dispersant which, by making avoid the formation of sludges, prevents the formation of the cells of differential aeration corrosion.

3° In order to avoid the fouling of cooling surfaces:

- through biological control;
- through partial sidestream filtration (of the flow recirculated in "sliding current");
- through the use of dispersing agents.

4° In order to avoid the proliferation of Legionnella bacteria, following measures can be taken:

- the use of demisters as effective as possible;
- bactericidal treatments adapted and made reliable.

5° In order to avoid pollution due to the blowdown:

- the use of non-toxic chemicals for the treatment.

6.4 Basic Principles of the Treatments

In order to have some ideas concerning numbers, we will suppose that the make-up water contains following elements, in the proportions indicated:

Cl^-	Chlorides, ppm	30
Mg^{++}	Magnesium hardness (in ppm of $CaCO_3$)	30
Ca^{++}	Calcium hardness (in ppm of $CaCO_3$)	70
MA	Methyl orange alkalinity (in ppm of $CaCO_3$)	115
SO_4^{2-}	Sulphates, ppm	20
SiO_2	Silica, ppm	15
pH		7,2

By reasons that will come to light later, we will suppose that the pH of the recirculation water is of 6.5 to 8.9.

6.4.1 Scale

Whenever the concentration of a salt is superior to its limit of solubility, it settles in the form of scale, in general, on a hot surface (as the solubility is lower in a hot environment than in a cold one). The salts producing scale that is found the most often in cooling waters are: silica, calcium sulphate, calcium phosphate and calcium carbonate.

As for the first two salts, there are the empirical calculation rules, which are set forth as follows:

1° *Silica will settle if its concentration in the recirculation water is superior to 50* x *(pH — 3.5).*

For a pH of 6.5, this corresponds to a concentration of 150 ppm, and the permissible concentration is of 150/15 = 10 cycles. For a pH of 8.9, the concentration would be of 270 ppm and the permissible concentration would be of 270/15 = 18 cycles. In all the cases, the maximum permissible concentration of silica in the recirculation water is divided by the silica concentration in the make-up water.

2° *Calcium sulphate or gypsum, will precipitate if the value of 0.4 (THCa* x *SO_4) exceeds 500,000 for the water with high pH.*

At pH 8.9, this value is of 560 and the permissible concentration is 500,000/560 = 893 cycles. For the water with a low pH, the value of 0.4 (THCa, in ppm Ca CO_3) (ppm SO_4) + PA should not exceed 500,000. At the pH of 6.5, this value is of 3780 and the permissible concentration is of 500,000/3780 = 144 cycles.

These theoretical values of the limit concentration ratios should not make forget that the gypsum concentration is a limiting element for the concentration ratio, as the gypsum provokes a physical degradation of the concrete which constitute the common material for the most considerable cooling towers.

In all the cases, additional information should be obtained from the manufacturer of the concrete.

The reason for the calculation of these cycles consists in the following:

Rule: THE HIGHER THE CYCLES OF CONCENTRATION, THE LESS THE COST OF THE WATER SUPPLY AND ITS TREATMENT, SINCE THERE IS LESS BLOWDOWN TO BE CARRIED OUT.

The conditions in which the calcium carbonate (that is to say, the main element generating scale) precipitates are more complex. This precipitation depends on the calcium hardness, the total alkalinity, the total quantity of dissolved solids, the temperature and the pH.

The tendency of water to precipitate calcium carbonate is defined by an index called Langelier saturation index, or by an index of stability, called Ryznar stability index, which will be described in details in the chapter dedicated to the scale.

6.4.2 Corrosion

Corrosion is caused by a cell constituted by two different metals in contact, or by two dissimilar regions of the same metal (called anodic and cathodic regions) surrounded with water. An electric current leaves the anodic region of the metal, crosses the water until the cathodic region, and crosses the metal to return to the anodic region.

Corrosion occurs at the point where the current leaves the metal (anode). The only means to stop this kind of corrosion is to stop the current flow. A process to stop this flow consists in coating the metal with an insulating material, in order to form an electrical barrier.

A layer of scale strongly adhesive to the metal can form such a barrier. If the stability index of the recirculation water is inferior to 6.0, there will be the scale formation, which will help preventing the corrosion. If the stability index of the water in circulation is superior to 6.0, there will be no formation of scale and there will be the possibility of corrosion.

Rule: ALL THE NATURAL WATERS ARE EITHER CORROSIVE WITH REGARD TO THE MARBLE, OR GENERATE SCALE.

At first sight, it would seem that corrosion could be avoided, by fixing the stability index to a value which would provoke the deposit of a very fine layer of scale (controlled scaling). However, this is not practical, given the the influence of temperature on the stability index: scale will be formed at the hot end of a heat exchanger, while the cold end would be attacked.

A certain number of chemicals have been tested in order to coat steel, or allow the steel forming an insulating film which is so thin that it can not affect the heat exchange. These materials are called corrosion inhibitors. Some of them are used for insulating the cathodic regions, the others - for coating the anodic regions. Among them, we can cite:

Cathodic	*Anodic*
Polyphosphates	Orthophosphates
Zinc	
Molybdates	Nitrites
Polysilicates	Ferrocyanides
Orthosilicates	
phosphonates	

The purpose of this discussion is to make understand that there are, to simplify, two methods of efficient treatment:

- The first rule consists in **rendering the recirculation water corrosive, deliberately**, so as to avoid the scale, and then adding a corrosion inhibitor in order to avoid corrosion.

This first method, the simplest one for rendering the recirculation water corrosive, consists in adding enough acid (usually the sulphuric acid) in order to lead the pH to about 6.5. The value of 6.5 is chosen, because at a pH very inferior to it, corrosion inhibitors become inefficient. The experience has proved that it is necessary to add about 1 ppm of concentrated sulphuric acid in order to eliminate every ppm of the total alkalinity (expressed in $CaCO_3$) in the feed water.

Now that we have rendered the water corrosive, it is necessary to choose a corrosion inhibitor or a mixture in order to ensure the protection of the cathodic and anodic regions.

At present, the chromates zinc mixtures, very toxic, are no more in use. The polyphosphates zinc mixture has been used less and less, because of the action of phosphates on the eutrophication of lakes.

The criterion for the choice of an inhibitor is first of all its non-toxicity for the environment, and its efficiency is examined in the second place.

Currently, the most commonly used mixtures are the zinc-phosphonate mixtures or carboxyphosphonate alone.

- The other method consists in putting ourselves deliberately in the scaling conditions, by operating at a pH called free pH, whose value is situtated between 8.3 and 8.8.

Operation at an alkaline pH limits naturally the necessary concentrations of corrosion inhibitor. This second method can imply the use of the sulphuric acid in order to adjust the pH or rather the MA (methyl orange alkalinity) at an optimum level determined using the objective concentration ratio.

The necessary quantity of corrosion inhibitor will be close to that which was eliminated during the blowdown, increased by some % necessary for keeping the inhibitor film on the metallic surfaces.

Free pH corresponds to the pH of the water equilibrium after disappearance of the balancing carbon dioxide, given the physical-chemical composition of the water.

In order to avoid the scale deposit, it is indispensable to introduce a product meant to prevent or retard the formation of deposit. It is a matter of a scaling inhibitor.

There are two families: phosphonates and polyacrylates.

They have the effect of retarding the precipitation of calcium salts and rendering them less scale-forming. We will see them again in the chapter dedicated to scale and its treatment (Chapter).

6.4.3 Clogging/Fouling

Clogging (fouling) is the deposit of substances, which are usually suspended, as opposed to scaling that means the deposit of substances, which are usually dissolved.

The suspended polluting matters contained in the recirculating water can be issued from many sources.

Fouling in a plate heat exchanger

These can be the particles of sludge (silt) issued from the make-up water, or the airborne dust entering into the tower, or wood fibres. They also may come from the products used for the treatment, or may result from the presence of micro-organisms in the circuit.

These microorganisms include the bacteria, algae and fungi.

Bacteria are found wherever in a cooling system. The most embarrassing ones are the bacteria forming the organic sludge and those reducing the sulphates.

The bacteria forming the sludge are protected from external attacks by a thick gelatinous layer. When these organisms settle on the cooling surfaces, this layer can greatly hinder the heat transmission. It can also prevent the corrosion inhibitors from entering into contact with the metallic surfaces, and thus, provoke the corrosion of the surfaces.

Most of the bacteria encountered in the cooling systems are aerobic, and they consume carbonaceous compounds and oxygen with production of carbon dioxide.

However, the sulphate-reducing bacteria are fed on soluble sulphates and the product of their elimination is the hydrogen sulphide: they thus generate a corrosive environment for steel.

The oxygen is toxic for the sulphate-reducing bacteria. Thus, theoretically, they are not likely to develop in the cooling water in circulation. However, certain species form a protective layer, and the others live under the protective layer formed by the sludge-producing bacteria.

One of the most frequently used biocides is chlorine. The method recommended is to continuously add chlorine to the water leaving the tower, so that the water returned contains a residual chlorine rate of 0.1 to 0.2 ppm.

There is no simple method to calculate the necessary concentration of chlorine. The appropriate quanity should be determined empirically for each system.

Shock treatments with more significant concentration may be necessary.

The chlorine can eliminate the bacteria only if it enters into the contact with them. The protective layer formed by the bacteria producing organic sludge prevents this contact. In the case of a problem, the tower can be energetically treated in the form of periodical shocks with 20 to 30 ppm of the quaternary ammonium salt, in the same time with the chlorine treatment as it has been described above.

The function of the quaternary ammonium salt is to dissolve and to detach the protective layer and to allow the chlorine affect the bacteria. The quaternary ammonium salts have germicidal properties as well.

The presence of reducing agents in the recirculation water renders the quantity of chlorine required for certain systems too large to be applicable.

In this case, it is necessary to use other biocides. There are a large number of such biocides, and it is difficult to anticipate how each biocide will react in a particular situation concerned. Tests are usually necessary. The most commonly used active materials are the isothiazolones, the 2.2-dibromo-3-nitrilopropionamide which is degraded at an alkaline pH (DBNPA), the tetrahydroxyphosphonium sulphate (THPS) whose toxicity in relation to the environment is very low.

Each of these active materials has:

- a given action spectrum;
- a speed of action as regards the effect on such or such microorganism;
- a specific " toxicity profile" which regroups the information such as the lethal dose, the service life depending on the pH, the degradability, etc.;
- a given cost.

The choice of an ideal biocide should be made depending on the characteristics specific to each circuit, depending on the quality of make-up water, the place of discharge, the holding time, etc.

Algae are microscopic plants with chlorophyll. They are usually green and need the sunlight in order to develop.

They can reduce in the same time the air flow and the water flow. Their weight alone can constitute a significant increase of the weight that the construction must bear. The dead algae are an additional load of suspended solids in the recirculation water.

The key of the algae control problem consists in the fact that they need the sunlight in order to multiply. Thus, the algae do not present a serious problem in the coolers with closed platform. In the case of open towers, if there is an algae problem, an algicide should be used.

Fungi are also plants, but without chlorophyll. They grow on a large number of materials, including wood, where they are fed, and then they break up. The common method used for preventing this consists in using, for the construction, treated wood or materials other than wood.

In the case where the clogging/fouling is not due to a biological phenomenon, it is of course not preventable by biocides. It is therefore necessary to use the products that ensure keeping dispersed the insoluble elements which cause fouling.

They include polyacrylic acids and their salts. Their function is to destroy the electrostatic charges and to conserve the suspended colloid or to let it settle in the tank of the cooling tower. Depending on the product used, a concentration of the dispersing agent varying from 10 to 30 ppm in the recirculation water is usually sufficient.

Problems due to the proliferation of microorganisms in the cooling circuits and their solutions make the subject of a special chapter (Chapter 9).

The last important subject is the pollution treatment. A part, sometimes considerable, of the wastewater of a plant is constituted by the blowdown of the cooling systems. The products/materials used for the treatment are found again in the blowdown.

The discharge of the blowdown into natural environment can only be considered in the case of non-toxicity of the products used. Anyway, the discharge should be carried out by taking into account the relevant legislation.

We will thus have the interest to choose, among the products for treatment, the least toxic ones or those which are the easiest to detoxify, or to use the synergie that exists between different products, so as to always use the minimal dose.

7

Deposits and Their Treatment

We have seen that the deposits found in a cooling system are of two types:

— hard and adhesive deposits which all are scale of mineral origin;
— clogging deposits which can be powdery and can give silt or sludge, when the circulation velocity of the water is not sufficient for keeping them suspended. They are mostly of the mineral origin. These deposits can be more or less adhesive and more or less soft, but, when they are of the biological origin, they can take any kind of appearance.

In fact, they are often constituted by a mixture of the elements of organic, mineral and biological origin.

7.l Scale

Scale is mineral crystalline deposits caused by the formation of crystals in a supersaturated solution. They are usually dense, well cohesive and highly adhesive to substrates.

7.1.1 Formation of Scale: Solubility Product

A slightly soluble electrolyte is dissolved in water until a solution saturated by its ions is formed. Let us consider a saturated calcium solition with the solid calcium sulphate.

If a given quantity of solid calcium sulphate dissolves, the number of calcium ions and of sulphate ions formed is the same:

$$CaSO_4 \text{ (solid)} \leftrightarrows Ca^{++} + SO_4^{--} \text{ (saturated solution)}$$

By applying the law of chemical equilibrium to this system, we obtain following mathematical expression:

$$[Ca^{++}]\,[SO_4^{--}] / [CaSO_4] = K_e$$

Given the fact that the calcium sulphate concentration remains basically constant, we have:

$$[Ca^{++}]\,[SO_4^{--}] = Ke\,[CaSO_4] = K_{Ps},$$

where K_{Ps} is called *the solubility product.*

If the chemical equation for the deposit formation involves the same ion several times, the concentration of this ion is raised to the corresponding power in the solubility product equation. Thus, tricalcium phosphate decomposes (dissolves) according to the following equation:

$$Ca_3(PO_4)^2 \leftrightarrows 3\ Ca^{++} + 2\ PO_4^{3-}$$

and the relevant solubility product is:

$$[Ca^{++}]^3 \times [PO_4^{3-}]^2 = K_{Ps}$$

7.1.2 Influence of Temperature on the Solubility Product

The solubility product for the calcium sulphate is of:

— 900,000 at 25 °C (expressed in ppm);
— 840,000 at 60 °C (expressed in ppm);
— 710,000 at 80 °C (expressed in ppm).

Since the solubility product decreases when temperature increases, we see that the calcium sulphate, like most of the calcium salts, is less soluble in hot water than in cold water. Thus, the deposit will be formed on the hot wall, to which it will adhere, and not within the colder water.

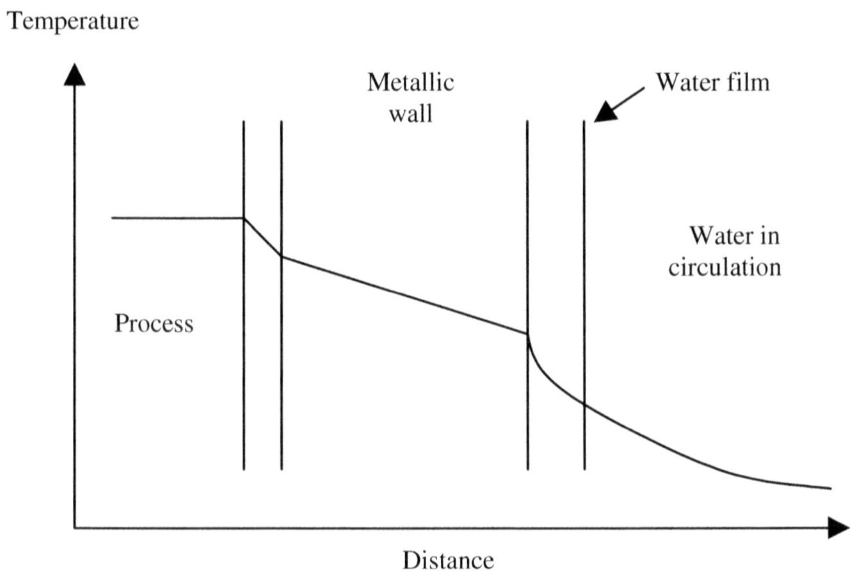

Temperature gradient

Temperature in question is therefore not the temperature of the water mass in circulation, but the temperature of the water in the (form of) thin film which is in immediate contact with the wall of the tube.

In practice, there is no formation of scale, if we have:

$$[Ca^{++}]\ [SO_4^{--}] < 500{,}000,$$

where the concentrations of SO_4^{--} and Ca^{++} are expressed in ppm.

If we maintain the product of the calcium concentration (or 0.4 x hardness caused by calcium) and the sulphate concentration in the circulating water below 500,000, then no scale of calcium sulphate (or gypsum) will be formed on the cooling system surfaces.

7.1.3 Influence of pH on the Solubility Product

Temperature and concentration are not the only variables affecting the solubility product. Certain components generating scale are sensitive to the presence of other ions, in particular to the presence of hydrogen or hydroxyl ions, which are connected with pH.

For example, silica is more soluble in the alkaline water than in the acid water. Again, we have an empirical equation which anticipates the silica deposits and (which) is set forth as follows: silica forms a scale deposit, if its concentration in the cooling water exceeds 50 (pH — 3.5). For a pH of 6.5, we can assume: 50 (6.5 — 3.5) = 150 ppm of silica before scale will deposit.

The calcium phosphate is another product which is sensible to pH. Given the fact that the phosphate ion is rarely present in the natural water, but it is added as a treatment agent, its concentration does not vary as that of the calcium.

7.1.4 Calcium Carbonate

The conditions in which the calcium carbonate would form a scale deposit are more complex. This depends on the calcium hardness, the total alkalinity, the total quantity of dissolved solid matters, temperature and pH.

This problem was studied by Langelier.

It is a study of the carbonic acid equilibrium of calcareous water; he refers to all the equilibria that exist at a given temperature between the ions H^+, OH^-, CO_3^{--}, HCO_3^-, Ca^{++}, et H_2CO_3, which leads to establishing a relation between the pH, calcium hardness, alkalinity and temperature.

In order to avoid long calculations, various equations of calcium carbonate stability have been presented in the form of diagram like the Langelier nomograph shown in the following figure. The study realized by Langelier, presented in the form of graph in the following page, constitutes a method of prediction allowing know whether the water in question tends to dissolve or to precipitate scale. If the water precipitates scale, it is scale-forming; if, on the contrary, it dissolves scale, it is corrosive.

The Langelier saturation index is equal to the difference between the current pH of the water and the pH of saturation (pH, at which it would be neither scale-forming, nor corrosive). This pH of saturation pH_s is obtained using the values of temperature, calcium hardness, methyl orange alkalinity (MA or m-alk) and the concentration of total dissolved solids.

In practice:

- we transfer vertically the calcium hardness expressed in ppm of $CaCO_3$ to the diagonal of the pCa and we read horizontally pCa;
- we transfer vertically the total alkalinity expressed in ppm of $CaCO_3$ to the diagonal of the pMA (french: pTAC) and we read horizontally pMA (pTAC);
- we report vertically the total solids expressed in ppm to the isothermal curve "constant c" of the temperature desired, and we read horizontally the value of constant c on the scale c.

The pH of saturation is equal to the sum:

$$pCa + pMA + C = pH_s$$

and we obtain:

$$\text{Langelier saturation index} = pH - pH_s$$

Langelier saturation index

diagonale des pCa
diagonale des pTAC

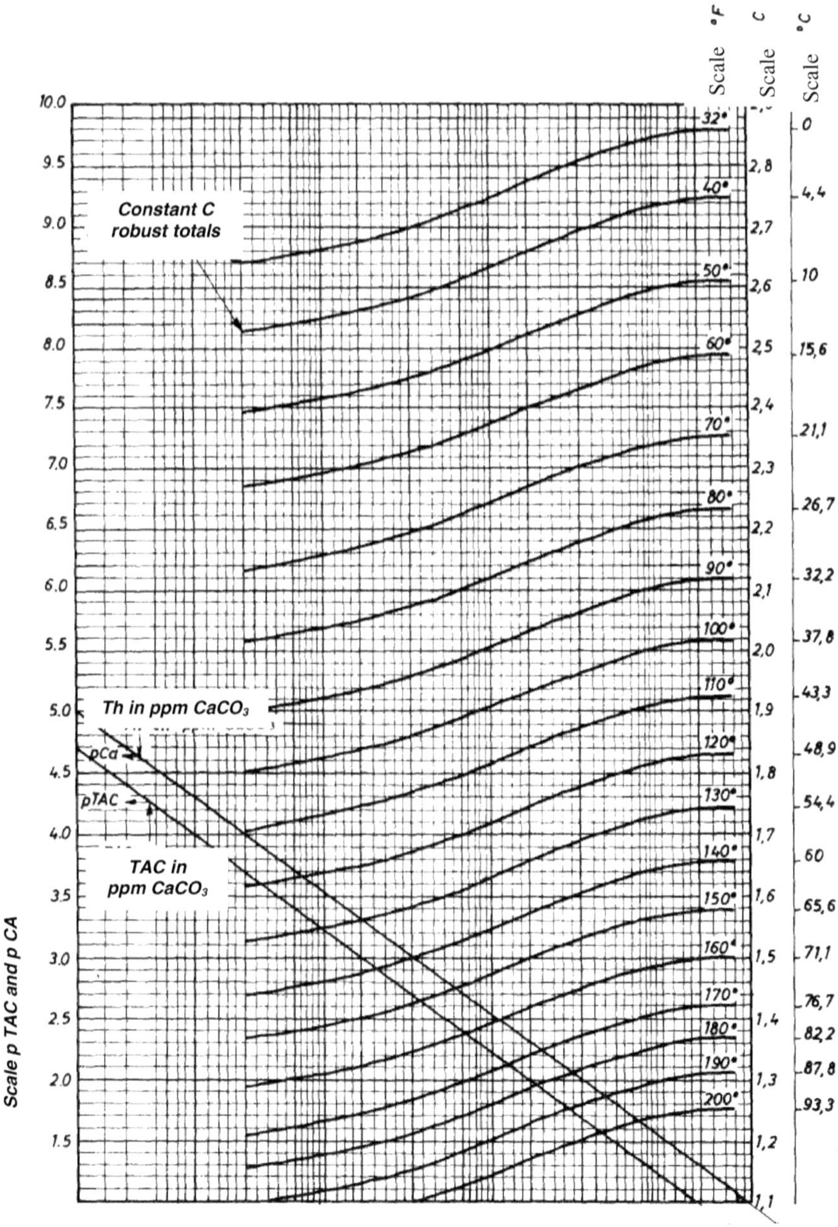
Scale °F
Scale C
Scale °C
Constant C
robust totals
Th in ppm $CaCO_3$
pCa
pTAC
TAC in
ppm $CaCO_3$
Scale p TAC and p CA
10.0
9.5
9.0
8.5
8.0
7.5
7.0
6.5
6.0
5.5
5.0
4.5
4.0
3.5
3.0
2.5
2.0
1.5
32°
40°
50°
60°
70°
80°
90°
100°
110°
120°
130°
140°
150°
160°
170°
180°
190°
200°
2,8
2,7
2,6
2,5
2,4
2,3
2,2
2,1
2,0
1,9
1,8
1,7
1,6
1,5
1,4
1,3
1,2
1,1
0
4,4
10
15,6
21,1
26,7
32,2
37,8
43,3
48,9
54,4
60
65,6
71,1
76,7
82,2
87,8
93,3

Interpretation:

pH — pHs > 0	scale-forming water
pH = pHs	"neutral" water
pH — pHs < 0	corrosive water

Another similar method of evaluation to resolve the same problem is the Ryznar stability index.

We have:

$$\text{stability index} = 2\ pH_s - pH$$

as pH_s is the pH of Langelier saturation, the interpretation is as follows:

Ryznar stability index	Water tendency
4to5	Intense scale-forming
5to6	Light scale-forming
6to7	Light scale-forming to light corrosion
>7	Corrosion

The stability indices of the water circulating at 38° and at pH 8.9 are the following for different cycles of concentration:

Cycles	Stability index	Tendency to scale deposit ($CaCO_3$) without treatment
1	6.56	No scale deposit
2	5.40	Light scale deposit
3	4.70	Moderate scale deposit
4	4.20	Significant scale deposit
5	3.70	Very significant scale deposit

This table shows that water leads to the scale formation at any - or almost any – cycle of concentration.

Other indices, like that of Puckorius and the SAL factor of L. Duvivier, are also used.

► *Puckorius index IP*

$$IP = 2\,pH_s - pH_e$$

$$pH_e = 1.485 \log (MA) + 4.54.$$

The analysis of this index is similar to that of the Ryznar index.

► *Factor SAL*

In order to characterize the scale-forming tendency of the water in question and to fix the limits of concentration to respect, we can also use the SAL factor developed by L. Duvivier, which corresponds to the following relation:

$$SAL = [MA] \times [MA]2 / f$$

f is a coefficient that depends on the ionic strength of the water. The SAL factor depends on temperature. The limit value is determined in/at a pilot plant, in normal operation conditions of the circuit.

A scale deposit that is found less frequently is that of magnesium silicate, which is formed in two stages: first, the magnesium hydroxide precipitates, then it reacts with dissolved colloidal silica, which leads to a dense scale deposit, which is difficult to remove. A solution to the problem is to prevent the formation of the magnesium hydroxide. A method for the determination of pH of magnesium hydroxide is indicated in the supplement enclosed. If the value of this pH is inferior to the pH of the cooling water in circulation, a scale deposit based on magnesium silicate can be produced.

An anhydrous scale deposit can be produced on the extremely hot surfaces of the pipes (that is to say, at more than 120 °C) even for the solubility products as low as $(Ca^{++})\,(SO_4^{--}) = 50{,}000$. This is due to a design problem: in this case, air-cooled batteries or heat exchangers should preferably be used, rather than water (type) coolers.

A comparison of solubilities of different calcareous scale deposits is presented below.

There are scale deposits of zinc phosphate as well. They are usually due to a bad application of the water treatment.

So-called "scale deposits based on phosphate of iron and iron oxide" are indeed the products of clogging/fouling, but they are normally not adhesive.

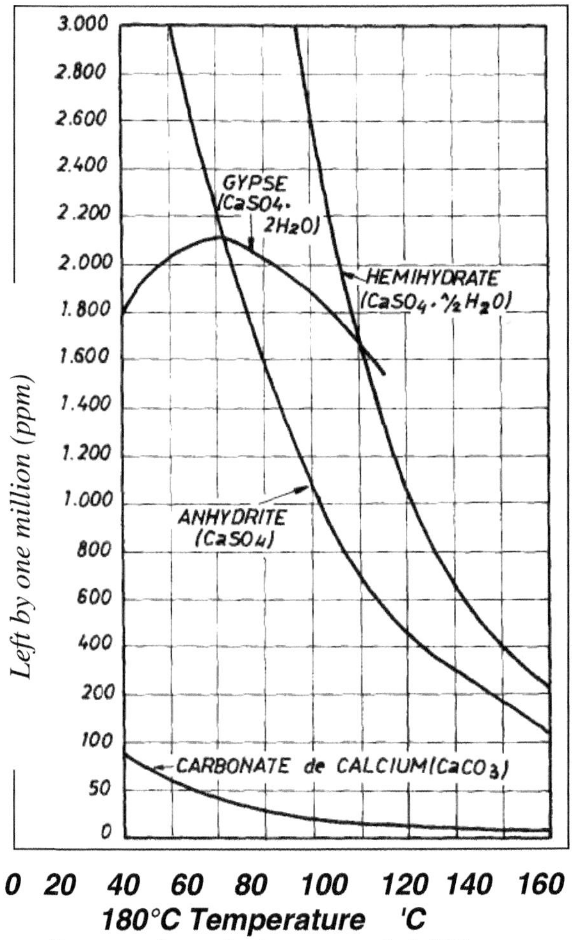

Comparison between solubilities of scale-forming calcium salts

7.2 Scale Prevention

Two types of process can be considered:

— Subtractive processes: removal of one or several scale-forming ions;
— Additive processes: addition of the substances which will bring under control the solubility product, or alter the process of crystal growth, or disperse the crystals formed.

7.2.1 Subtractive Processes

This question was referred to in the first part of this book, concerning the pretreatment where we have decribed total and partial softening processes: the one using ion exchange resins and the lime softening.

These methods imply investments including the expenses necessary for the process, which are accompanied by costs related to labour. Their implementation is to be made after a techno-economic study.

The criterion related to the availibility of make-up water ressource may be a decisive one.

7.2.2 Additive Processes

These are the processes used in the open recirculating cooling circuits with tower in almost all the cases, although the make-up water undergoes a pretreatment.

7.2.2.1 Solubilization by Acidification

The acidification is one of the oldest processes used for preventing the scale formation. The calcium bicarbonate is transformed into sulphate according to the following reaction:

$$Ca\,(HCO_3)_2 + H_2SO_4 \rightarrow CaSO_4 + 2\,H_2O + 2\,CO_2$$

As the calcium sulphate is much more soluble than the carbonate, this treatment is efficient even in a plant that has to operate with a large number of cycles. If the concentration of sulphates in the make-up water is already high, the addition of sulphuric acid can make exceed the solubility product of calcium sulphate. In this case, it may become necessary to use hydrochloric acid, or another acid. (Figure below).

The cooling circuits with tower operate more and more at free pH (pH between 8.0 and 8.5), some of them operate in neutral medium (pH between 7.0 and 7.5) and some others - in acid medium (pH between 6.5 and 7.0). This problem will be discussed later, in the chapter concerning the corrosion control.

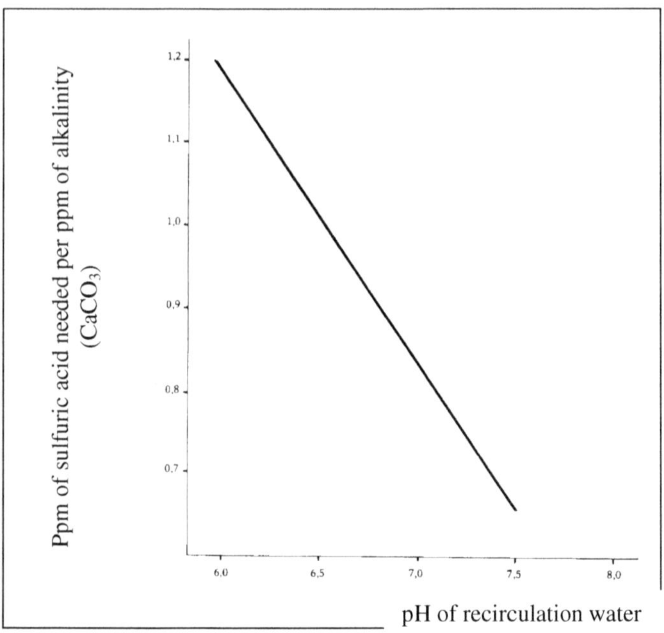

7.2.2.2 Solubilization by Chelating Agents

Solubilization by chelating agents like EDTA and NTA (ethylene diamine tetra-acetic acid and nitrilotriacetic acid) is very efficient, mainly at alkaline pH, but these chemicals must be used in stoichiometric doses (ratios), and thus are no longer used in cooling water; however, they are still used in boilers.

7.2.2.3 Addition of Scale-Preventive Dispersing Agents

At the difference of previously-discussed processes which limit the concentrations of scale-forming ions, the addition of scale-preventive agents allows these ions concentrating beyond the limits of natural precipitation, by delaying the insolubilization and deposit of $CaCO_3$ through a physical-chemical action on the crystallization process, while the concentration of the agent is still well sub-stoichiometric.

They do not prevent the precipitation of calcium carbonate in thermodynamic sense of the term, since they do not react with the latter. The concentrations applied, of the order of some ppm, are too low for the reaction to take place. The precipitation is only delayed.

Many researchers have studied the parameters of precipitation and modifications of the crystalline structure in the presence of scale-preventive agents.

We can draw following general conclusions from their works:

- crystals of calcium carbonate are deformed and have less tendency to stick together;
- time necessary for the precipitation increases with the increase in concentration of the additive;
- the quantity of the precipitated calcium carbonate decreases when the concentration of additive increases.

The mechanism of action of these inhibitors is not very well known. The phenomenon is more likely to have an action on the crystalline growth than on the process of nucleation. The crystalline growth is usually caused either by the diffusion of ions or molecules on the surface with continuous deposition, or by the fixation of ions or molecules on a layer adsorbed to the surface of the crystal. If, at a given moment of the crystal formation, the additives are adsorbed to the surface, they can reduce and modify the process of growth.

Anionic organic polymers of small molecular weight (1,000 to 10,000) are also likely to disperse certain crystals already formed. In fact, in an aqueous solution, they are ionized to acquire a negative charge. By coating the "colloidal" particles of $CaCO_3$, they create on their surface a zone of negative charge. The colloids, charged in this way, repel one another and they do not adhere neither to one another nor to the walls.

Scale inhibitors:

They are divided into two groups. Molecules based on phosphorus and carboxylic organic polymers.

► *Molecules based on phosphorus*

Inorganic polyphosphates

These are prepared by dehydration of appropriate orthophosphates. Their general structure is characterized by a repetitive linkage - P - O - P -.

Indeed, it is a matter of the first scale inhibitors which were used in the industrial field. The most well-known among them is sodium hexametaphosphate.

Among the main disadvantages related to their use, we can note: their very low stability in aqueous medium, due to the insuffcient solidity of the linkage P-O-P; their hydrolysis is particularly dependent on temperature and pH. They are degraded into orthophosphates, which have no stabilizing property.

Moreover, the orthophosphates cause the formation of deposits of calcium phosphates, which, although not scale-forming, reduce considerably the heat exchange.

Finally, the orthophosphates are nutrients for microorganisms which can cause an excessive proliferation of algae and bacteria both in the cooling system and in the natural environment where the blowdowns are discharged (phenomenon of eutrophication); their stabilizing action is limited to the maximum concentrations of bicarbonates of 4 to 5 meq/1;

For all these reasons, polyphosphates are practically no longer used in water treatment except for their corrosion inhibiting action. They are studied in the chapter concerning corrosion.

Phosphonic acids and their salts

These agents act, apparently, rather by inhibition of crystal formation than by real sequestration. In fact, at the difference of the chelating agents, they are effective at the doses well inferior to those determined by the stoichiometry. The inhibition is produced as soon as these compounds at very low concentration are adsorbed to the surface of the scale-forming crystal germs, thus preventing them from swelling, while scale remains in pseudosolution. They act through the "threshold effect".

These are organic products obtained through synthesis. They are usually prepared by reaction of phosphorous acid with acid anhydrides and/or acid chlorides, in particular, those of acetic, propionic, butyric and valeric acids. Instead of using phosphorous acid, we can make react directly the phophorus trichloride with a carboxylic acid. Thus, the 1-hydroxy-ethylidene1.1-diphosphonic acid (HEDP) results from the synthesis of the phosphorus trichloride and the acetic acid in an aqueous medium.

The phosphonates, whose molecule contains one or several atoms of nitrogen, are relatively sensitive to chlorine. Thus, there is an incompatibility between the use of nitrogenous phosphonic acids and continuous injection of Javel water.

The HEDP is used frequently, because of its stability in the systems using chlorine. Another phosphonic acid, the amino trimethylene phosphonic acid (AMP) is also used, but it is sensitive to oxidizing agents, including chlorine.

HEDP

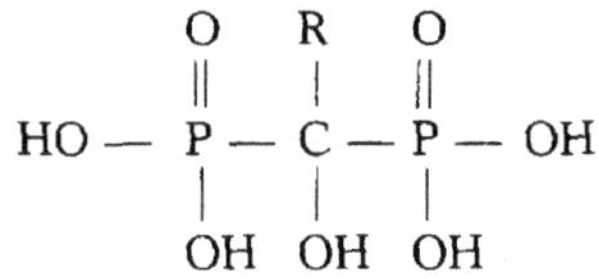

AMP

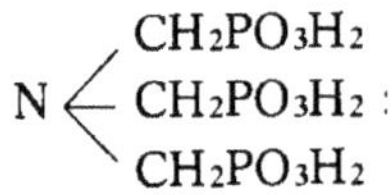

Phosphonic acids and their salts can sequester the copper. When this metal and its alloys (Admiralty brass, for example) are present in the circuit, it is necessary to associate to them a corrosion inhibitor for copper and copper alloys such as benzotriazole or tolyltriazole.

In some conditions, the phosphonic acids can lose their scale-preventive function as regards the alkaline-earths. This phenomenon, called "turbidity zone", is shown by the figure below. At higher doses of phosphonate, there is an appearance of a turbidity which results from the precipitation of calcium phosphonate.

Even higher doses, close to that of stoichiometry, redissolve the precipitate, but it does not present any economic interest. The phosphonate concentration, at which the precipitate appears, depends on the conditions of pH and calcium concentration. A high pH and a very high calcium concentration can lead to the precipitation at very low concentrations of phosphonates.

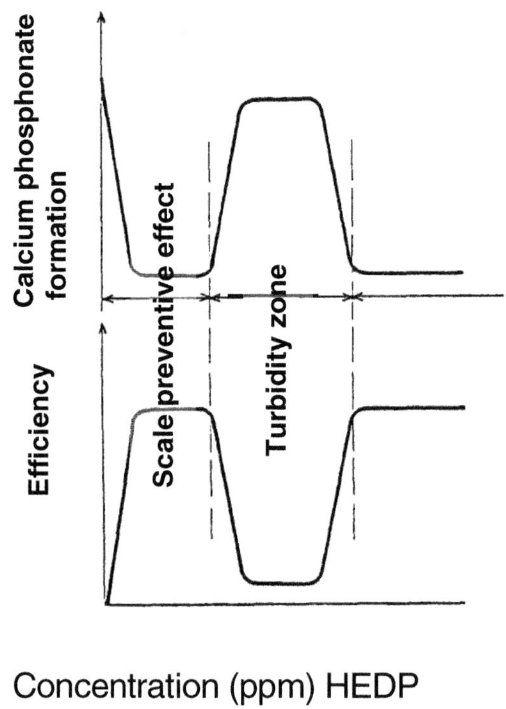

Turbidity zone

► *Organic polymers*

Organic polymers used for the scale-preventive treatment of water all belong to the family of polycarboxylic acids with small molecular weight and their salts. Their molecular weight is relatively small: 1,000 to 10,000. It is a matter of, in most cases, the derivatives of polyacrylic, polymetacrylic and polymaleic acids.

Thus, the polyacrylic acid is manufactured using the acrylic acid dissolved in water; we add potassium persulfate as catalyst. The temperature must be controlled and maintained constant, depending on the polymer that we desire to obtain. The polymerization reaction is stopped due to the introduction of a chain stopper. The preparation is then cooled and the polyacrylic acid is slightly neutralized by caustic soda, potash or ammonia. Finally, in order to avoid any odour, we remove remaining monomers by oxidizing them with hydrogen peroxide. The quantity of the catalyst, the temperature and the moment where the chain stopper is introduced, determine the molecular weight and the structure of the polymer.

We use particularly the polyacrylates of the following formula:

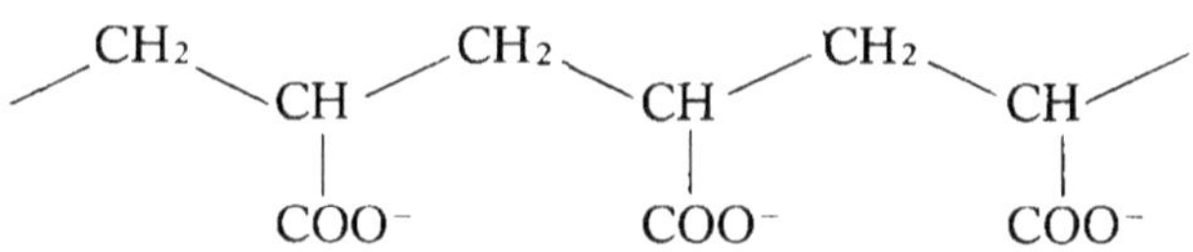

or the polymethacrylates:

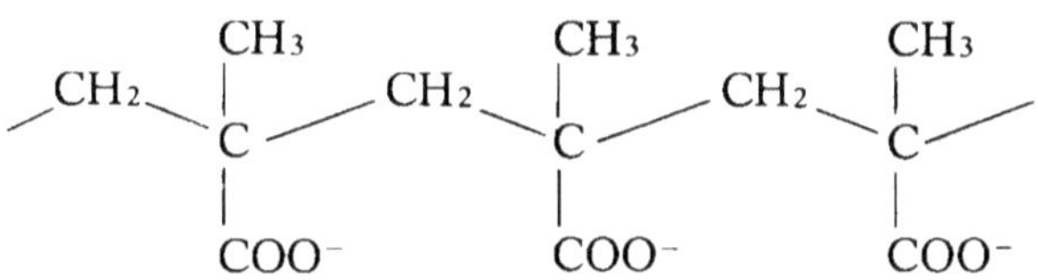

Most of these polymers currently available on the market, resist, without notable degradation, to the temperatures of 250 to 300°C. They are perfectly stable to chlorine. Thus, there is no incompatibility between their use and continuous injection of sodium hypochlorite.

Phosphonates and the mixtures of phosphonates and polymers are in general better scale-preventers than the polymers alone. They allow maintaining a Ryznar index of the order of 3.50.

7.3 Fouling Deposits

Fouling deposits are constituted by solid mineral or organic substances, initially suspended, which form usually not thick and voluminous deposits, hindering the water circulation, reducing the heat transfer, and which provide an excellent substrate for microbial growth and reduce the effectiveness of anticorrosion treatments. These substances have a natural origin, but they can also be caused by process leakages.

They can be:

a) **of natural origin**:

Issued from water

(1) Silt and sludge
(2) Organic matters
(3) Dissolved minerals
(4) Micro and macro-organisms

Issued from air

(1) Gas
(2) Dust and ground dust
(3) Organic matters (vegetation)
(4) Microorganisms
(5) Insects, etc.

b) **of artificial origin**

Issued from water

(1) Matters carried after passage in a clarifier
(2) Dissolved or suspended organic matters
(3) Phosphates
(4) Detergents
(5) Sewer effluents

Issued from air

(1) Organic gas
(2) Hydrogen sulphide
(3) Sulfur dioxide (SO_2)
(4) Ammonia

Issued from the system

(1) Corrosion products
(2) Corrosion inhibitors and their reactions
(3) Process leakages

7.3.1 Principle of the Deposit Formation

7.3.1.1 Coarse Particles

The formation of deposit is due to the action of gravity on the particle which has a density greater than the liquid and falls according to Stokes' law (see pages 7 and 8). According to this law, the settling velocity in water at rest is quite variable, depending on the nature of the suspended particle and particularly on its diameter.

Diameter of the particule		Time for fall of 1 metre
10 mm	Gravel	1 second
1 mm	coarse sand	10 seconds
0.1 mm	fine sand	2 minutes
0.01 mm = 10μm	Silt	2 hours
1μm	Bacteria	4 days
0.1μm	fine clay	2 ans
0.01μm	Colloids	10 ans

As the water is in circulation, a deposit is produced, as for the coarsest particules, only when the liquid flow velocity is inferior to the falling velocity (of the particle); thus, it is in the zones of little agitation that deposits will be formed.

7.3.1.2 Fine Particles

At diameters below 1 μm, the Stokes' law is no longer applicable, as the particles become places where electrostatic phenomena take place, thus keeping them indefinitely in suspension.

Let us examine in details one of these fine particles, whose surface is very significant in relation to its weight.

The negative ions in the medium have tendancy to be adsorbed to the surface of the particle (figure a), which causes an unstable state, since free negative charges can not exist in the nature.

Thus, a layer of positive ions is established, more or less, on the surface (figure b). These positive ions, called *Stern layer,* compact layer or rigid layer, neutralize only partially the negative charges.

Further in the solution, there is (a layer with) a mixture of positive and negative charges, with a surplus of positive charges sufficient for the complete neutralization (figure c) of the above-mentioned negative charges.

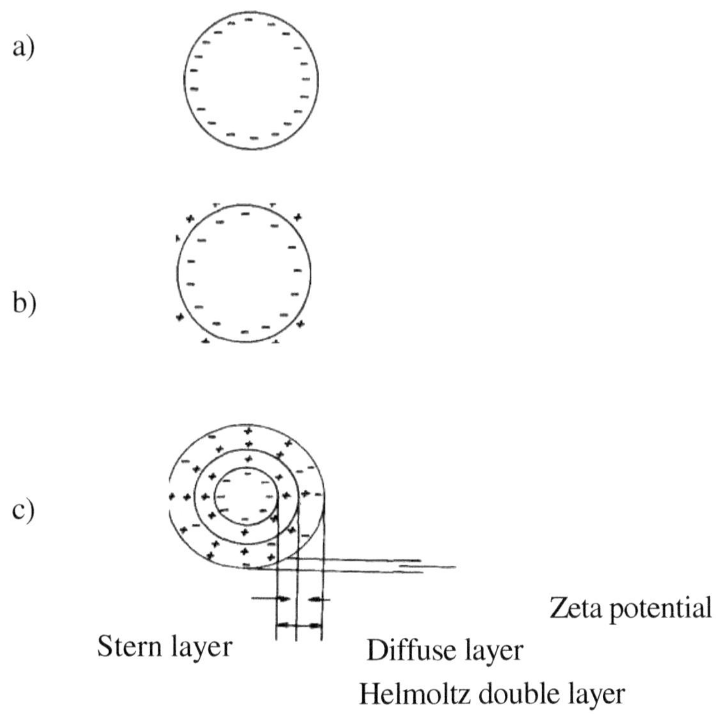

The positive charges are extremely mobile. The zone between the Stern layer and the electrically neutral solution is known as Gouy layer, *or diffuse layer.* The layer encompassing these two layers is called "Helmholtz double layer".

The electric charges cause mutual repulsion of the individual particles, producing what is called the Brownian motion. Thus, the individual particles can neither deposit nor enter into collision with each other to stick together in larger masses which could settle.

When a charged particle moves in relation to the surrounding medium, some amount of liquid moves with it as well. Between the liquid carried with the particle and the mass of liquid, there exists a boundary called *shear plane (or slipping plane).* The potential (in millivolts) of the Gouy layer at the level of the shear plane is the *zeta potential.*

The zeta potential, that is to say the charges usually negative (electrons) due to superficial ionization of practically all the particles in aqueous suspension, is one of the most important natural forces. Its prime purpose is to create the initial repulsion necessary for maintaining the individuality of the cells suspended in water, whether it be in the plant kingdom or animal kingdom. If this repulsive charge is reduced below its normal value (— 15 to — 20 millivolts), there would be coagulation and cell death result.

What is good for living organisms is not always good for the cooling systems with water circulation. In all the cases where water is in contact with steel surfaces designed for heat exchange, iron atoms tend to pass into solution. This is not corrosion, but it is a phenomenon of dissolution.

In order to pass into solution, the metal must be ionized. Thus, we have a surface covered by a positive charge, and when a negatively charged particle in suspension is found close to this surface charged positively, it is attracted by the electrostatic effect.

Finally, it will lead to a considerable layer reducing the heat exchange, and the equipment becomes fouled or clogged.

Fouling is the deposit of usually suspended material, as opposed to scaling, which is the deposit of normally dissolved material.

As iron is more soluble in water than copper is, the positive charge on its surface is greater. Thus, the steel tube bundles become fouled more quickly than the tube bundles made of "admiralty" brass, copper, zinc or tin.

Other causes of deposit should not be left without mentioning; these are the deposits formed by bacterial colonies, algae, bacteria and fungi, which are formed by cellular proliferation.

They often result in organic formation: mycelium of fungi, fibrous algae, but, even when the cellular formations have isolated cells, they are connected between them by a mucus, often thick, which is a source of significant fouling. The latter point is related to the chapter of microbiology and will be considered later.

7.4 Prevention of Clogging Deposits

7.4.1 Physical Means

Although they necessitate greater investment than chemical treatments, they are sometimes indispensable and often useful.

In the cooling systems with water recirculation, a sidestream filtration of a part of the recirculated water is very efficient. It allows removing clogging substances coming from make-up water, from the air and polluting products of the process. There are many types of efficient filters used for the removal of solid matters, as well as of oil. In general, a small quantity (usually from 3 to 8 %) of the recirculation water is filtered. Due to this, suspended particles are reduced to a minimum and the concentration of clogging substances is restricted at least as regards the larger particles.

Filters, and even screens, at the location of the air intake of cooling towers can reduce the concentration of airborne fouling substances, in particular, in air scrubbing systems used in the textile and tobacco industries. The screens must be used whenever problems are caused by various bird and insect debris.

7.4.2 Chemical Means

They constitute the essential of the deposit prevention and at present we have an entire range of products:

► *Polymers*

- Natural organic polymers were the first to be used; these were tannins, alginates and starchy substances; sometimes they gave quite good results, but only when they were used at relatively high concentrations: 50 to 200 ppm or even more;
- Acrylic polymers of low molecular weight, which are efficient at much lower doses, have now supplanted them. These are, in general, polyacrylates most often presented in the form of sodium salts. They have been discussed in the previous paragraph.

These low molecular weight polymers have a particularly significant anti-deposit effect. The commercially available polymers have a molecular weight inferior to 10,000.

They act in the same time as complexing agents for polyvalent ions (complexing, but not sequestering) and as dispersants by adsorption; however, as their molecular weight is sufficiently low, they do not have any coagulating effect.

These polymers have a dispersing effect in the same time a scale-preventing effect by deforming the crystal lattice of the precipitates. Moreover, they prevent deposit.

Other synthetic polymers act in the same manner and they are usually based on maleic anhydride.

► *Polyphosphates*
Polyphosphates having different chain lengths have been used since many years in the treatment of ferric sludge and the removal of calcium scales.

Classically, they are used in the treatment of domestic hot water, alone or in association with polysilicates.

However, polyphosphates have the property of hydrolyzing slowly, and thus losing their efficiency to some extent. This question will be treated in the chapter dedicated to corrosion, as these products are used especially as anti-corrosion agents.

► *Polynaphtylméthanes sulfonates* (PNMS)
These are anionic dispersants having in their molecules several naphtylsulfonic nuclei linked between them by methylene bridges. They only have a dispersing effect with no foaming effect. They are active at all the pH encountered in the cooling water.

They maintain in suspension the deposit formers, which are then removed during blowdown. They also have a remarkable effect of fluidization of mineral suspensions of clay and sludge, and allow removing sludge much more rich in dry matter in the pretreatment facilties.

They are not much used in cooling circuits.

► *Surfactants*
Surfactants are molecules having a heavy hydrophobic chain or group and much lower mass solubilizing function. Their main property or role is to lower the surface tension of the water and to act as an emulsifying agent.

They are used for emulsifying oily substances, fouling or hindering the evaporation of the water, which may be found either due to the use of polluted water, or due to leakages in the equipment. They are also efficient for dispersing sludge of microbiological origin. Moreover, the reduction of the surface tension potentializes the effect of biocides used. Some of these chemical compounds can form foam in the zones of high turbulence.

7.5 Anti-deposit Treatment Implementation

Knowledge of the existing treatment products is of course indispensable, but not sufficient for choosig an anti-deposit treatment. In fact, deposit can be provoked by or come from the elements found in the make-up water. In any case, it is necessary to know the nature of the deposit.

Following table presents the typical analysis of the deposits and we give an interpretation for each case:

Typical analysis of deposits and source of clogging elements (on dry sample: 24 hours at 110 °C)

Test	A	B	C
Ash	30%	17%	55 %
Al_2O_3	2 %	35 %	12 %
SiO_2	1%	13 %	8 %
CaO	10 %	4%	1 %
Fe2O3	30%	25 %	23 %
P_2O_5	—	4%	1 %
ZnO	25 %	—	—
CO_2	2%	—	—
SO_3	—	2%	—
Total	100 %	100 %	100 %

Sample A comes from a circuit with cooling tower of a large chemical plant using a zinc/organic compound formulation for the control of corrosion and deposit. The clogging products are emphasized by a significant loss of ignition and the presence of iron oxide and zinc oxide. In this case, the corrosion inhibitor itself precipitates and appears in the form of zinc oxide and loss of ignition. This deposit can be removed by means of a detergent and an acid scrubbing, but in order to prevent its reappearance, it is necessary to change the corrosion inhibitor.

Sample B is a deposit coming from a direct once-through cooling circuit used in a large refinery. The case is quite frequent with many cooling systems using river water. High concentrations of aluminium and silica oxides correspond to alluvia of the river water. The consequence of this deposit is corrosion of equipement, thus formation of iron oxide. Other components emphasized by the analysis come from pollutants contained in the river water and trapped in the deposit. Given the fact that no corrosion inhibitor is used, the fouling substances are most easily removed by means of a fluidifier of sludge and silt. An acid wash may also be effective to a certain extent. However, a frequently-used preventive method is continuous addition of a dispersing agent.

Sample C is a deposit coming from a cooling system supplied by river water in a refinery. The analysis, again, emphasizes the problem of clogging and trapping of sludge (Al_2O_3, SiO_2) in the deposit. A periodic treatment with chlorine has not been sufficient for avoiding these deposits.

Installation of a sidestream filter could effectively clean the system. It is also necessary to set up an effective treatment with dispersant and biocide so as to solve all the problems of the system.

► *System operation and characteristics*

The operation of the cooling system and its specific characteristics must continually be supervised. Opening of heat exchangers and examination of the installations during programmed stops allow checking the absence of deposit. If necessary, a sampling must be made for the purpose of analysis.

► *Velocity*

Every cooling system depends on uniform and constant flow of water on the heat exchanging surfaces. Since fouling substances are usually suspended in the water, they will deposit in zones of lower velocity, particularly in regions where velocity varies abruptly, for example, in heat exchange tanks, on the walls of tube bundle shells, in tower basins or in air condensers.

► *Temperature*

Clogging (or fouling) risks increase when temperature rises. This is due to the "burning" effect on the surfaces where the clogging products settle, the tendency to scale deposit, the increase of the rate of corrosion, faster reactions and a loss of activity of some dispersing agents.

► *Type of equipment*

The type of equipment used can contribute to the increase of clogging. The heat exchanger tubes, which depass the tube plate, for example, can lead to a faster fouling. The bimetallic couples of tubes made of "admiralty" brass, and the tube plates or deflection plates made of steel can cause clogging due to the corrosion products.

► *Source and treatment of water*

The water used in a cooling circuit varies depending on its origin or the treatment that it has undergone. The water changes depending on the atmospheric conditions, that is to say, its turbidity, salinity, hardness, etc. The quality of the treated water can change during its clarification, depending on the chemical products added. Sometimes, solid particles can be carried (phenomenon of postflocculation).

► *Observations and human control*

The operation of cooling systems must seriously be monitored. A bad use of the equipment, for example, the reduction of the flow through heat exchangers, can cause a serious fouling or clogging. A correctly programmed starting of the circuit leads to a good control of deposits.

The clogging can be controlled by:

1° removal of clogging substances after their formation and a regular maintenance program;
2° a periodic acid scrubbing, aiming to remove the clogging substances;
3° most often, by means of a prevention program.

A combination of prevention and cleaning is often used, when the prevention program is not entirely efficient.

7.6 Control and Monitoring of Scaling and Fouling

The control of the quality of heat transfer and its evolution is indissociable from the evaluation of the quality of water conditioning implemented on a cooling circuit. The efficiency of a production unit depends on the quality of heat transfer.

The control can be realized in several manners:

7.6.1 Monitoring of the Deposit Thickness on a Test Surface

- Setting up of test pieces at a location corresponding to the highest temperature;
- Taking out these test pieces after 4, 6 or 8 weeks;
- Examination and analysis of organic or mineral deposit;
- Drying at 110 °C during 1 hour;
- Measurement of the deposit thickness through variation of magnetic induction, if the support is magnetic, and through eddy current, if the support is not magnetic.

The test pieces used for the corrosion monitoring can be adapted to the scaling monitoring.

The nature of the deposit and its thickness, depending on its exposure time, inform us on the type of treatment dysfunction and the corrections to be made.

This comparative control in time has the disadvantage of not exposing the test piece to a heat exchange.

7.6.2 Monitoring of the Head Loss of a Test Exchanger

The monitoring, after cleaning, of the head loss of an exchanger whose stable water flow is known, and its conditions of heat transfer in terms of difficulties, inform us qualitatively on the evolution of fouling.

This method is particularly adapted to the fouling of biological origin.

7.6.3 Monitoring of a Test Exchanger of the Circuit

This is a very important measure, as it expresses the efficiency of the production unit. The quality of heat exchange is represented by the transfer coefficient:

$$K = q / S \times \Delta^{TM}$$

where : $\Delta^{TM} = [(T2 - t1) - (T1 - t2)] / \mathrm{Log}_e [(T2 - t1) / (T1 - t2)]$

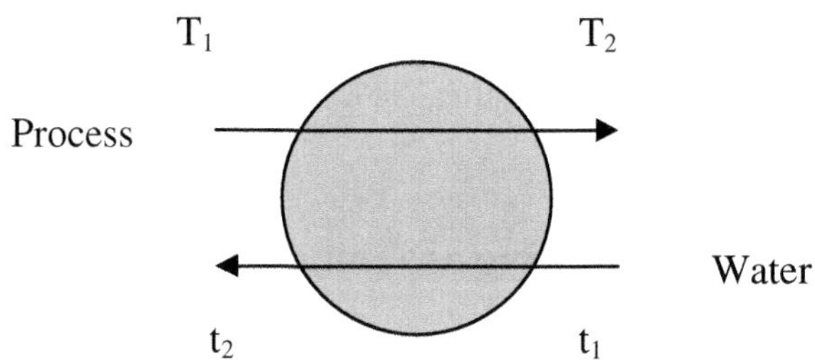

K = transfer coefficient (kcal/$m^2 \cdot h \cdot °C$)
q = thermal load (kcal$\cdot h^{-1}$)
S = exchange surface (m^2)
t_1 = inlet cooling water temperature (°C)
t_2 = outlet cooling water temperature (°C)
T_1 = process inlet temperature (°C)
T_2 = process oulet temperature (°C)
with: $q = Q \times (t_2 - t_1)$,
where Q is cooling water flow.

The measurement and systematic control of the heat transfer coefficient can thus be carried out, using the knowledge of six values: four temperatures and two flows. Here, again, we must have reliable measurements.

These methods of control of scaling and fouling are still associated with the methods of analytical monitoring.

a) *On the recirculation water*:
- Comparison of concentration ratios: TH, Ca (calcium), SO_4 (sulphates), Cl^- (chlorides), when it is possible.
- Bacteria count.

b) *On the deposits collected:*
- Quantitative and qualitative measuring of the elements.
- Analysis using X-ray spectroscopy, which allows the identification of the mineral compounds and their form of crystallisation.

8

Corrosion and Its Treatment

► *Generalities*

The prevention of corrosion in the open recirculating cooling systems is a field, in which enormous progress has been made during the last twenty years.

It has widely been demonstrated that the total costs of anticorrosion treatment are very low compared to the savings realized by the reduction or absence of operating losses due to the stops and prolongation of the service time of the equipement.

In fact, the anticorrosion treatments are as important as the scale-preventing treatments, since it is the service life of the cooling unit itself that is at stake. The corrosion control ensures the durability of the equipment.

► *La corrosion*

In general, corrosion can be defined as the progressive destruction of the construction materials by their surroundings, which includes at first the **electrochemical corrosion of metals**, which we will mainly discuss in this chapter, but this includes also mechanical erosions: erosion by solid matters in movement and erosion under the influence of fluids at high speed (cavitation). It includes microbiological attacks of the materials as well.

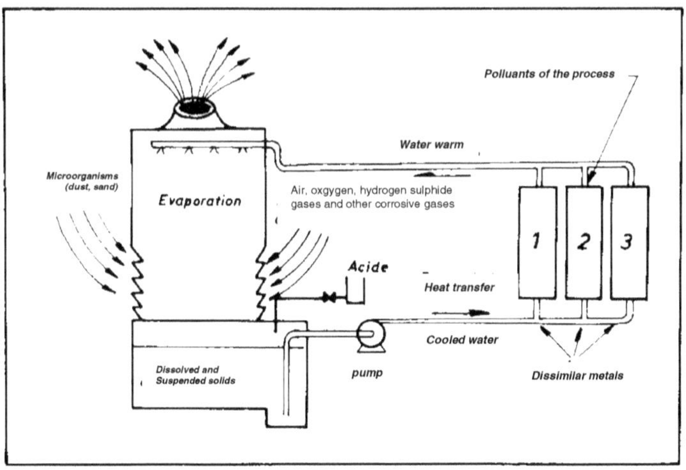

1. Steel heat exchanger
2. Aluminium heat exchanger
3. Condenser made of bronze (copper, zinc, tin.....

Corrosion factors in cooling towers

8.1 Theory of the Electrochemical Corrosion of Metals

Electrochemical corrosion is the oxidation of a metal by its surroundings; we will consider here as surroundings only water, and then, there always exists an electrochemical phenomenon. We will see that corrosion is always connected with a phenomenon of heterogeneity.

8.1.1 Galvanic Corrosion

Experience shows that a metal, and particularly an alloy, always behaves as a heterogeneous material, due to the impurities, its crystallinity, its mechanical strain, etc. Therefore, it exists, between neighbouring points of the surface, low but real potential differences, and as the metal is in contact with impure water which is conductor, an electric current runs between the metallic surfaces through a solution. There is corrosion at the place where the current leaves the metal and passes into the solution. This surface is called anode. The surface where the current returns to the metal is called cathode. In addition to the conducting solution (electrolyte) and two electrodes (anode and cathode), there must be an electron pathway to complete the electric circuit. The structure of the metal itselft may provide this pathway for electrons, or the circuit may be completed by a physical contact between the metals of the anode and the cathode in the case of two different metals.

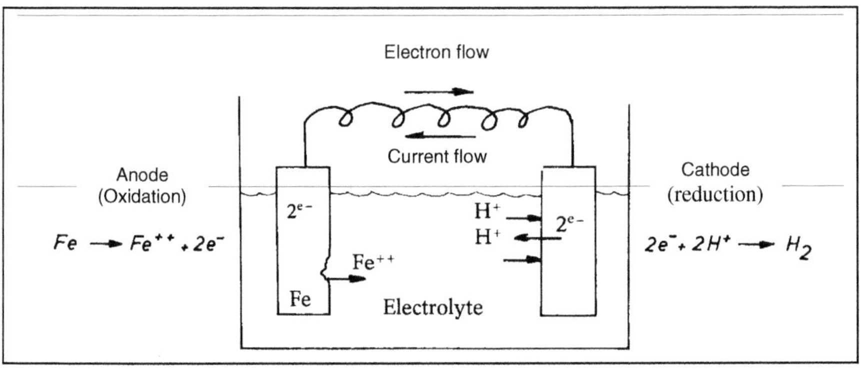

The simplest case is that of the galvanic corrosion of two dissimilar metals. Well-known examples are a steel tube screwed in a bronze valve, and tubes made of cuprous alloy such as admiralty brass, beaded on the tube plate made of black steel of a condenser or of a heat exchanger.

A neutral atom of iron loses two electrons and becomes an iron ion Fe^{++} in solution in the electrolyte. The electrons flow in the opposite direction to the electric current, towards the bronze cathode and return in the electrolyte. They are combined with two hydrogen ions, charged positively, to form one molecule of hydrogen gas.

If all that occur is summed up in this very simplified diagram, corrosion would terminate once the system reached equilibrium. This condition is called passivation. However, other reactions occur as well. Removal of the hydrogen ions from the electrolyte at the cathode (producing hydrogen molecules) liberates a surplus of hydroxyl ions (OH^-). These latter react with iron ions to form the ferrous hydroxide:

$$Fe^{++} + 2\,OH^- \rightarrow . \ Fe\,(OH)_2$$

The ferrous hydroxide precipitates, allowing the corrosion reaction continue.

When there is oxygen in the water, as in the case of the circuits with tower, other reactions occur. The oxygen reacts with the ferrous hydroxide at the anode, to produce ferric hydroxide which is even more insoluble. This is what we commonly call "rust" and it gives reddish brown colour of the corroded steels. At the cathode, the hydrogen produced reacts with the oxygen, to form water. A more widespread description of galvanic corrosion in the presence of oxygen is shown in the figure of page 134, with the following secondary reactions:

$$Fe^{++} + 2\,OH^- \rightarrow Fe\,(OH)_2 \downarrow$$
$$O_2 + 2\,H_2O + 4\,e \rightarrow 4\,OH^-$$
$$4\,Fe\,(OH)_2 + O_2 + H_2O \rightarrow 4\,Fe\,(OH)_3 \downarrow$$
$$4\,H^+ + 4\,OH^- \rightarrow 4\,H_2O$$

When there is a couple of metals, it is possible to predict which one of the two will be attacked. If two of the metals in following table are coupled, the metal occupying the higher position will corrode (in most cases).

Metal	*Potential* (in volts in a standard solution of NaCl (by reference to a calomel electrode, with a decinormal solution of chloride of K)		
Magnesium	—1,73	Bronze	— 0,28
Zinc	—1,00	Copper	— 0,20
Aluminium	—0,85	Stainless steel (18-8)	— 0,15
Cadmium	—0,81	Monel	— 0,10
Iron	—0,63	Silver	— 0,08
Lead	—0,55	Nickel	— 0,07
Tin	—0,49	Inconel	— 0,04

NB: Different values are obtained with another electrode of reference, for example, an electrode with hydrogen.

As we have seen, it is not necessary that there are two dissimilar metals in the system. Many metals, in particular iron, can have at the same time the anodic regions and the cathodic regions. When they are submerged in an electrolyte, the current will run from the anodic region towards the cathodic region, and the corrosion will continue.

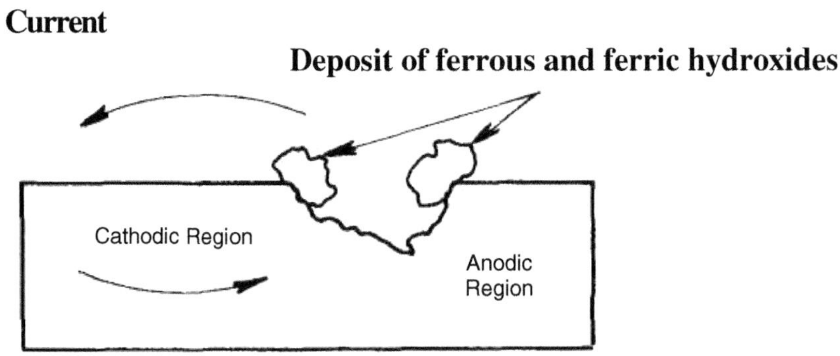

In practice, the anodic and cathodic regions are situated on the same part of metal, and the metal itself furnishes the pathway for electron return.

The differences of potential between the anodic and cathodic regions are caused by grain boundaries, direction of the grains, differences in size of the grains, differences in heat treatment, roughness of the metal surface, scratchings, differences in reaction to various stresses, deposits found there, etc.

8.1.2 Corrosion through Concentration and Differential Aeration Corrosion

Even a perfectly homogenous metal can corrode, if it is submerged in a non-homogenous liquid. The following figure represents the classical concept of a concentration cell.

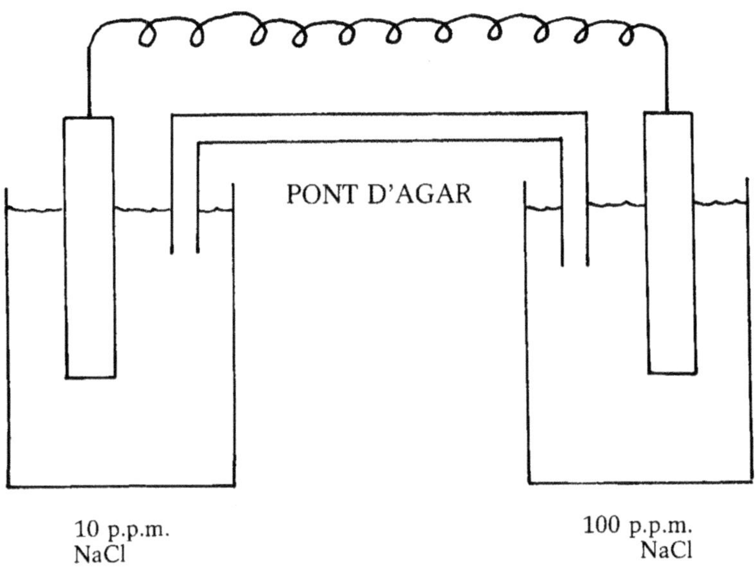

In general, the piece of metal immersed in the more dilute solution is anodic in relation to the piece of metal immersed in the more concentrated solution. In the above case, the difference of potential is of 0.06 volts, and the iron plunged in the solution of sodium chloride at 100 ppm corroded.

This type of corrosion is extremely widespread, because the chance to have difference of concentration in a point is as great at the side of the liquid, as that of the difference of structure at the side of the metal.

The most common cause is the differential oxygenation, as the water oxygenates through contact with air and then encounters less oxygenated humid walls, the least particle settled at the surface of the metal keeps between it and the metal a zone under-oxygenated in relation to the circulating liquid, hence the formation of a concentration cell. The iron will be dissolved in the under-oxygenated zone (anode) and the ions OH^- will be produced in the oxygenated zone (cathode). Even the differences of temperature will be at the origin of differences of potential (by different liquid densities, by different concentrations of an electrolyte, etc.).

8.1.3 Bacterial Corrosion

Without wishing to completely describe the corrosion of bacterial origin, which is treated in a separate chapter, let us give just one example of such corrosion, which is the most widespread in cooling water.

An anaerobic bacterium develops in the fouled parts of the equipment: Desulfovibrio desulfuricans, which uses sulphates in its metabolism as the source of oxygen, by producing the hydrogen sulphide, which attacks the steel with formation of black iron sulphide. We can highlight this attack of the sulphate-reducing bacteria by pouring on the black deposit the hydrochloric acid which will produce the odour of rotten eggs.

8.2 Types of Corrosion

It must immediately be noted that the corrosion is never avoided completely, but all the treatments have the purpose of bringing it to an acceptable value, and before all, to a uniform distribution.

Different types of corrosion can be found:

► *uniform corrosion*

It is characterized by a uniformly distributed dissolution of the metal, and it is the least hindering as far as it does not exceed an acceptable value (of the order of 50 μm per year for plain carbon steel). This is the target of all the treatments.

► *localized corrosion*

This corrosion is, on the contrary, much more desastrous and insidious, since for a very small total quantity of metal removed, the localization can lead to the weakening of the wall, fastly evolving until the formation of a hole.

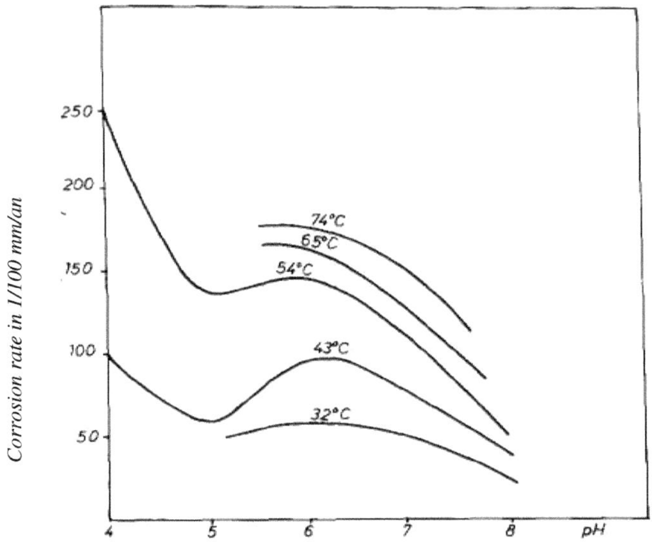

Rate of corrosion of the mild steel by non-treated cooling water in circulation

Corrosion by pitting consists in fast formation of small deep holes. This form of corrosion which is extremely localized is encountered most often when the temperature of the metal surface is high (superior to 60 °C). This type of attack would thus be related to the presence of microbubbles of air fixed on the wall. Pittings occur fastly when there is a contact between a steel wall and a copper wall. The copper becomes automatically the cathode and the iron - the anode (see page 134, the table concerning the potentials of metals). In this way, the copper is protected and it is the iron that is attacked.

Several conclusions can be made concerning this phenomenon: we should never use biocides based on copper in the equipment made of steel, galvanized iron or aluminium. Introduction of an iron piece in copper equipment will lead to fast destruction of the iron (large cathode, small anode). Pittings represent one of the most insidious and destructive forms of corrosion.

Tuberculation is an evolution of the pitting attack. The corrosion products form, above the hole, a tubercule of iron oxide having brown to black colour. We will see later that another type of tuberculation intervenes in the microbial corrosion; in that case, the tubercule contains iron sulphides.

Selective dissolving or separation is another aspect of the electrochemical corrosion. It occurs when one of the components of an alloy is attacked preferentially. It can be uniformly distributed or well localized.

This can be the dezincification of brass which lets subsist the porous copper having the red appearance of copper as opposed to the yellow colour of brass, the graphitization of cast iron or dissolving of the iron, which leaves a relatively non-resistant structure of graphite impregnated with iron oxide; the dealuminization of aluminium bronzes, which occurs especially in sea water.

Other forms of electrochemical corrosion can occur:

— attack at the water line, that is to say, at the zone of liquid-gas separation, which is a form of corrosion by differential oxygenation;
— cracks in metals under stress: mechanical stress between the crystals of the alloys are the zones much more sensitive to the corrosion.

This is a corrosion encountered less frequently in pure metals.

The erosion is, as opposed to the previously-mentioned corrosions caused only by the mechanical action, it is produced when a fluid hits a wall with high speed; there are, at the point of contact, very short alternations of high and low pressures; there are tractions of the protecting film and the metal, and the electrochemical corrosion can easily continue the attack. Too high a water circulation velocity, as well as certain operations of cleaning can cause significant erosion.

8.3 Methods of Corrosion Measurement

The evaluation of the level and the type of corrosion existing on different materials present in the cooling circuit constitutes the first index of the level of quality of a treatment. The pertinence of qualitative observations and confidence in quantitative values measured, allow necessary adjustments of treatment before the corrosion causes grave consequences for the durability of the working devices.

There are three methods of measurement:

- With corrosometer;
- With corrosion coupons;
- By examination of test pieces.

Whatever the method of measurement chosen, the observation point must be carefully determined, in order to avoid erroneous interpretations. In particular, the speed of the fluid and the temperature must be known and must correspond to the most extreme conditions of the circuit. The materials chosen must represent those present in the system.

8.3.1 Electric Measurement of Corrosion

The principle of the measurement by linear polarization consists in applying a small potential difference between two electrodes and to measure the resulting current, which is proportional to the speed of corrosion.

Devices allowing the instant measurement of corrosion through the observation of the polarization current are commercially available.

Besides the data on the corrosion (expressed in μm per year or in mpy: mils per year: 1 mpy = 25.7 μm per year), the devices provide an index of pitting (indication concerning the localized corrosion).

The number of electrodes (flat or cylindrical) is 2 or 3, and they must have the same shade/grade as the material that we wish to observe.

They allow not only the separate measurements with a portable device, but also a continuous recording of the corrosion current value.

These measurements require a rigorous, sometimes delicate, calibration between the corrosion current and the depth of the corrosion.

The results obtained must corroborate other means of control. The relative evolution of the values remains the main interest of this technique. The rapidity of the measurement is an additional advantage.

This system can serve as alarm in the case of dysfunctioning of the inhibitor dosage facilities. The corrosion of different materials can be controlled: carbon steel, copper, brass, stainless steel; the information on the pitting must be interpreted carefully.

Following figure shows a commercially-available device.

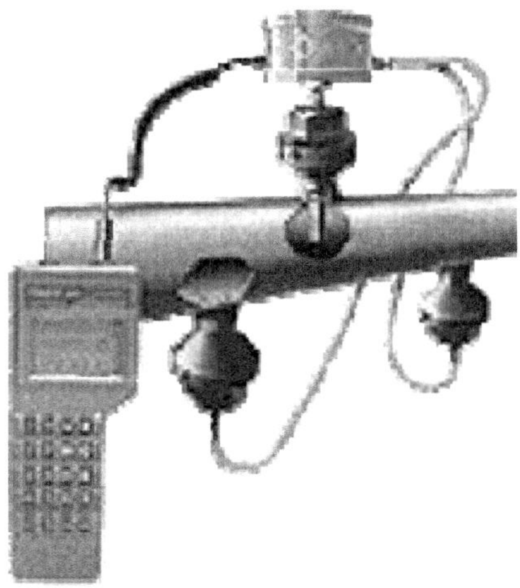

Electric measurement of corrosion (doc. Metal Samples)

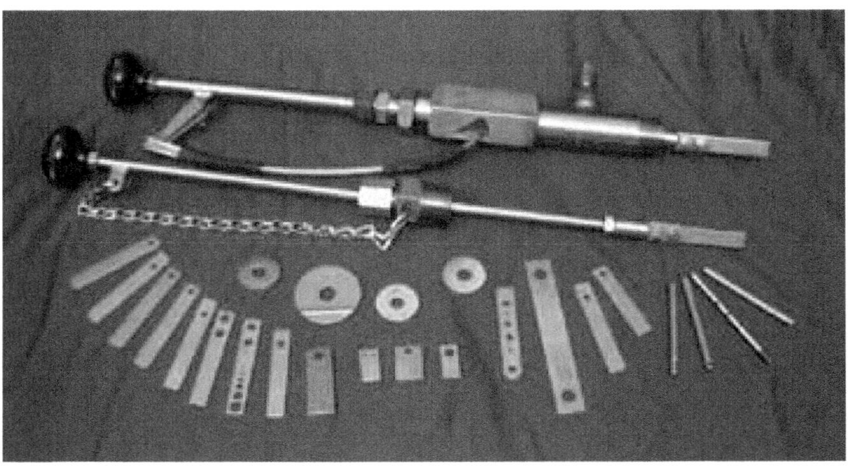

Examples of flat and cylindrical electric probes for corrosion control (doc Metal Samples)

8.3.2 Measurement by Corrosion Coupons

The metal coupons are immersed in water of the circuit to be controlled and are taken out periodically, in order to be weighed. The weight loss per unit of surface expresses the intensity of uniform corrosion.

The coupons can be placed in a by-pass loop (or corrosion rack) in the system. These facilities are usually made of PVC or plain carbon steel (ordinary steel). The following figure shows the system.

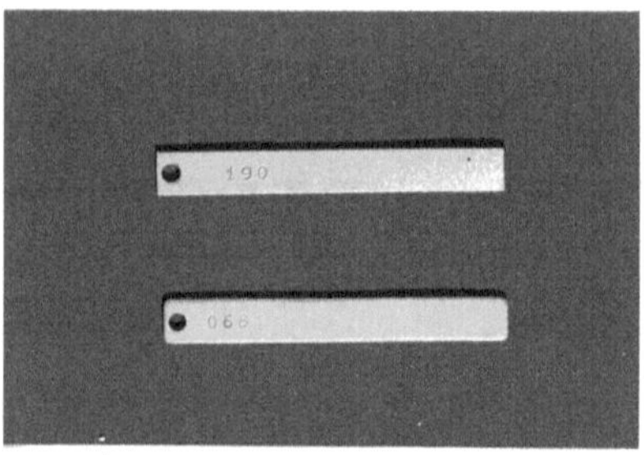

corrosion coupons

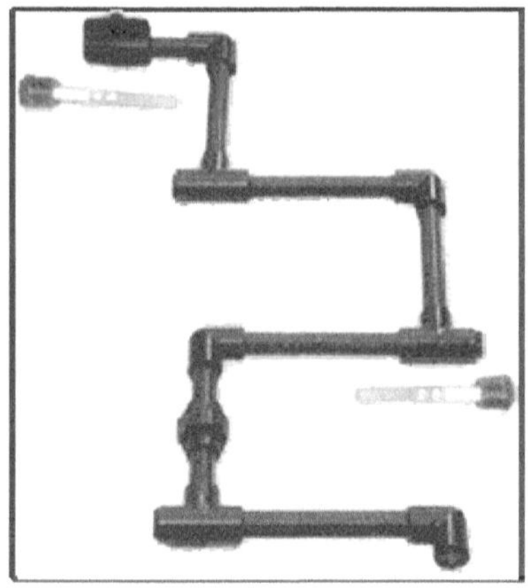

corrosion loop

A certain number of precautions should be taken during the design, preparation, arrangement and the examination of the corrosion coupons and their cleaning.

► *Design*

- Use of the metals identical to those present in a part of the system;
- The fixing of the coupons is made on a rod made of PVC or polyamide using a polyamide screw. This rod is fixed on the center of a stopper screwed on a PVC tube. Water circulates in parallel to the length of the coupon;
- The circulation velocity must be around 1 m/s, and the fixing should not create any turbulence;
- The coupons must remain constantly submerged even in the case of the interruption of circulation.

► *Preparation*

- Degreasing with acetone of each coupon before fixing it.
- The coupons in use have following dimensions, in general:

  - length : 3 inches;
  - width : 1/2 inch;
  - thickness : 1/16 inch.

Their surface: 3.38 square inches. The surface of each coupon is sand blasted.

There are coupons made of the most commonly used materials: carbon steel, admiralty brass, copper, stainless steel 304 L, 316 L, etc.

► *Cleaning*

- The exposure time must be at least 30 days before each weighing;
- A phase of degreasing with acetone is carried out prior to the weighing;
- Descaling in a 10% passive solution of hydrochloric acid during one minute;
- Brushing, then rincing in water distilled during some seconds and soaking in acetone;
- Drying at the ambient temperature during one hour and weighing to within mg.

► *Qualitative examination*

- After demounting the coupon: type of the deposit (mineral, organic, etc.);
- Sampling of the deposit for analysis;
- After descaling: microscope examination (magnification 50 times) in order to reveal pittings, crackings, etc., identification of the type of corrosion and counting of pittings per coupon face.

► *Interpretation of the results*

Corrosion in µm per year:

$$\frac{\Delta P \times 365}{d \times A \times N}$$

ΔP	weight loss (g)
d	metal density (g/cm^3)
A	exposed surface (mm^2)
N	exposure period (day)

Following values are considered to be good:

- non-alloy steel	< 50 µ/year
- stainless steel	< 5 µ/year
- nickel alloy of the type B or C	< 5 µ/year
- copper alloy	< 5 µ/year
- aluminium alloy	< 5 µ/year at pH 8

These values of uniform corrosion must be completed by the absence of localized corrosion (pitting).

8.3.3 Examination of the Test Pieces

We can consider easily demountable test pieces and examine them periodically.

It is also possible take samples of tube sections in the system or in an exchanger and carry out laboratory examinations, but the number and the choice are often limited.

We can also install a monotube test exchanger in order to simulate the industrial exchanger, and in this way, the problems of the skin temperature, the temperature gradient and scaling can be correctly replicated, and then the corrosion phenomena will also be replicated in appropriate conditions. It is necessary to continuously control the flow in such equipment, and this type of test lasts about 1 month. Although this process seems to be, in principle, the most attractive one, it is subject to caution because of the high temperature gradient, which exists at the inlet and the outlet of the tube, making the corrosion level abnormally high.

The evolution of the analysis of Fe^{3+}, Fe^{2+}, Cu^{2+} allows revealing the apparition of a corrosion phenomenon, and this is particularly true for the Cu^{2+}.

8.4 Spontaneous Corrosion Resistance of Metals

In the absence of treatment, all metals do not have the same resistance to corrosion, and it is necessary to know at least the major characteristics specific to each metal.

Ferrous metals have, in general, a bad resistance to corrosion, particularly in oxygenated medium, where there is a cathodic depolarization. The corrosion increases significantly at low pH, but an attack can also be produced in an alkaline medium (due to a weak amphoteric character). Despite all, this is the most commonly used metal due to its price, even if we must consider much larger thicknesses in order to take into account the wear due to the corrosion. Another disadvantage is the significant volume of corrosion products, which cause problems of fouling.

Cast iron contains more than 1.7 % of carbon, often up to 4 %. It is quite subjected to the graphitization, particularly in acid medium or very alkaline medium.

Stainless steels containing at least 12 % of chromium are self-passivated in the aerated water. The oxygen dissolves, re-forming the layer of passivating oxide, but until the temperature of 300 °C. Stainless steel is sensitive to chlorides, which cause pitting, and to corrosion in parts subjected to mechanical stress.

Galvanized iron has a resistance well better than that of the steel. However, beyond 65 °C it is sensitive to the alkalinity of the water. Morever, it is very sensitive to the copper ions, which give rise to pittings.

Copper is particularly resistant to the hard water, which provokes the formation of a protecting layer of carbonate. However, it is not very resistant to very soft water (pittings). It is resistant to the alkalinity, except the ammonia and the amines. A well-known defect of this metal is its lack of mechanical resistance.

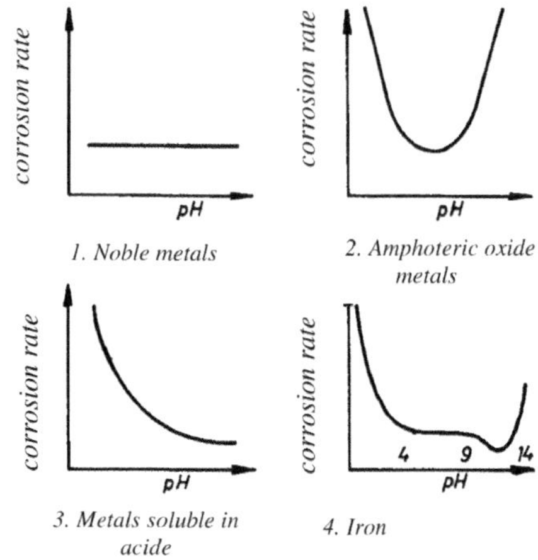

pH INFLUENCE ON CORROSION RATE

Brasses and bronzes are very resistant to the sea water, but they are subject to the dezincification, as soon as the concentration of zinc exceeds 15 %. This tendency is reduced by the addition of tin, antimony, phosphorus or arsenic.

The zinc is used especially for the galvanization.

Aluminium is also used due to its low density. As this metal is amphoteric, it is attacked by both the water with low pH and the water with high pH. The weakly acid water of pH 4.5 to 6 has practically no action; similarly, the aluminium is very resistant to the water of pH 6 to 8; from pH 8 to 9 there are risks of corrosion.

Beyond the pH 9 and below the pH 4.5 we have the risky zones; but, even in the zones of good stability, the aluminium is sensitive to the heavy metals, which considerably favour the corrosion attack. It is also to be noted that the impurities of this metal itself have a remarkable consequence on its resistance to corrosion (more than in other metals).

8.5 Anti-corrosion Treatments

We will describe here only the treatments against electrochemical corrosion, whereas the treatments against biological corrosion concern the paragraph dedicated to the microbiology.

Since an electrochemical cell is necessary for corrosion, the method to prevent corrosion consists in, logically, destroying this cell. A method to do so is to put a non-conducting barrier between the metal and the electrolyte.

The Langelier equation (previously discussed in the chapter concerning scale) necessarily implies that all the waters are either corrosive or scale-forming. It is often suggested that corrosion can be controlled by adjusting the stability index, so that a thin and adhesive layer of scale is settled on the metallic surface, insulating it from the electrolyte.

However, the stability index is very sensitive to temperature. If a stability index ensuring the protection is obtained using the *average temperature* of a cooling system, we state that only a part of the facilities of the cooling circuit is protected. There will be a scale deposit on the hot ends and the corrosion will attack the cold ends. Therefore, this method has no usefulness for the cooling systems. It is no longer used.

Currently, the basic principle of the cooling water treatment is: to render the recirculating water deliberately corrosive (a method among others to prevent scale), then add a combination of inhibitors in order to prevent the corrosion.

When we speak about the prevention of corrosion, it is a matter of bringing it to an acceptable value, because its complete prevention is an ideal that is never reached.

In general, we assume for corrosion supposed to be uniformly generalized (which is the first purpose to reach) the following figures:

- very low corrosion $< 20\mu$ per year;
- moderate corrosion 20 to 100 μ per year;
- unacceptable corrosion $> 150\mu$ per year.

Let us also remind that cooling towers operate at various pH ranges, which are the following:

pH 6 to 7 pH 7 to 8 pH 8 to 9

The choice of the pH range is made depending on the possible concentration ratio, the quality of the make-up water, the availability and the cost of the water.

Corrosion inhibitors are mineral or organic molecules, which, added to the water of the circuit, allow reducing considerably its corrosiveness.

There is a wide choice of corrosion inhibiting products. They are divided into three groups:

- *anodic inhibitors* form an extremely thin film on anodic zones, if they are used at very low dose, and on the entire surface – at much higher dose. The protecting films obtained are extremely thin. In order to be effective, they must be used in high doses, if they are used alone. Otherwise, they may, by leaving some zones non-protected, favour the attack by pitting;
- *cathodic inhibitors* produce a cathodic polarization which inhibits the reactions at the cathodes. These inhibitors are considered to be more reliable, as even in the case of insufficient dosing, they do not tend to cause any attack by pitting;
- *organic inhibitors* form the third type that is completely different from the above-mentioned two types in that the monomolecular protection is maintained on the metal by an adsorption process, without producing any notable modification of the metal's oxidation potential.

► *Synergy between inhibitors*

One of the most considerable advances in the field of anticorrosion treatment is to have observed that the combination of inhibitors, most often anodic and cathodic, not only had an action much more complete than each taken separately, but also that much lower doses of each were sufficient to obtain a satisfactory effect. Thus, the costs of treatment became interesting, besides a better efficiency obtained.

► *Description of the inhibitors*
Currently used inhibitors are the following:

Anodic	Cathodic	Organic
Orthophosphates	Polyphosphates	Phosphonates
Nitrites	Zinc	Thiazoles
Orthosilicates	Molybdates	Triazoles
	Polysilicates	Film-forming amines
		Tannins

Polyphosphates (cathodic)
These are the condensed forms of phosphoric acid used in the form of alkaline salt. The general formula is as follows:

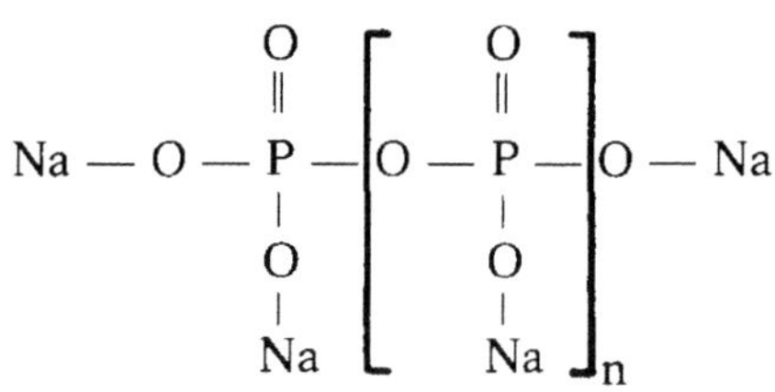

n is the degree of polymerization
n = 0 orthophosphate (will be discussed later)
it is an anodic inhibitor
n = 1 pyrophosphate
n = 2 tripolyphosphate
When n is superior to 3, the products have no longer the definite crystalline structure, but become vitreous. Among them, the most frequently used ones are:
n = 5 hexametaphosphate
n = 9 decaphosphate
They have following characters:

a) these are cathodic inhibitors forming a film containing calcium, iron and phosphate. They are considered to have no risk;
b) they necessitate a minimum of 100 ppm of calcium and precipitate beyond 400 ppm of calcium;
c) they are effective in the pH range of 6.5 to 7.5, in which their rate of hydrolysis is minimum. Their effectiveness is quite reduced in terms of total corrosion, but they avoid tuberculation;

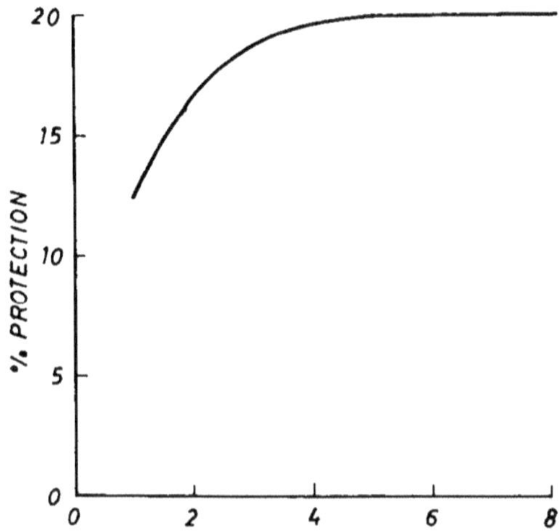

DAYS

CORROSION INHIBITION BY
2 ppm OF POLYPHOSPHATE

d) they do not react with reducing agents in the circulation water;
e) in general, they cause clogging and reduce the protection in the water with low velocity (below 0.3 metre per second). In the water with high velocity (beyond 3 metres per second), the film is easily damaged;
f) they constitute food for the algae growth in the cooling towers. When they are poured out into the system of the wastewater treatment, they can cause a proliferation of algae in the settling tanks. They can contribute to the eutrophication, but they can be removed by an appropriate treatment;

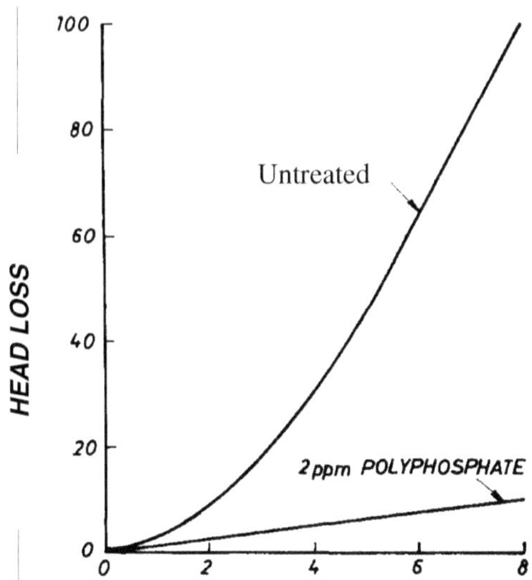

g) they do not protect copper and aluminium alloys. In fact, they attack these alloys, but this phenomenon can be corrected.

The main disadvantage of these products is their hydrolysis (or reversion) into orthophosphates, although the latter are anodic corrosion inhibitors.

Orthophosphates (anodic)

a) these are anodic inhibitors, and therefore, they can contribute to the pitting;

b) they must be used at high concentrations in order to have an efficient result;

c) they can cause scale deposit with calcium and clogging with iron, and therefore, they are used quite rarely.

Zinc

a) it contributes to clogging, in particular at the values of pH superior to 8.0 and at the concentrations superior to 3 ppm, although it is maintained in solution at a concentration of 5 ppm and at a pH of 6.5 – 6.7. The precipitation of the zinc hydroxide can be controlled through the formation of a complex with phosphonates.

CORROSION RATE OF RECIRCULATIN COOLING WATER TREATED WITH 60 PPM OF POLYPHOSPHATE (40 ppm of PO_4)

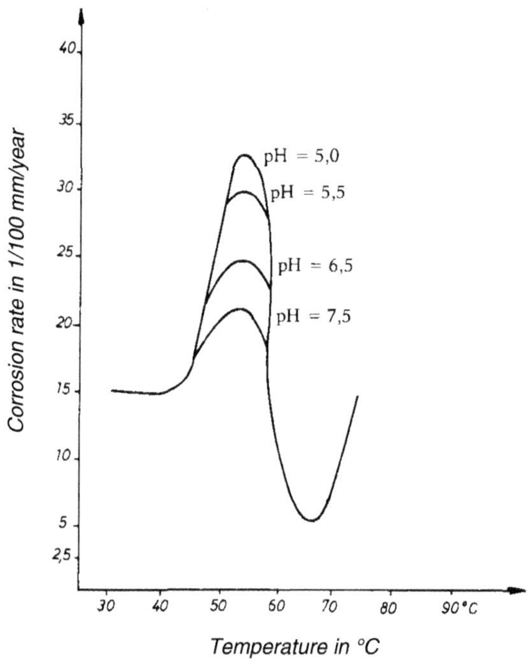

b) it constitutes a food for microorganisms at the concentrations inferior to 1 ppm, but at the concentrations superior to 3 ppm, it becomes toxic for the microorganisms present in any system of biological treatment of wastewater;

c) it is never used alone, but often combined with other inhibitors, as its film (although it forms quickly) is relatively non-resistant;

d) it is a cathodic inhibitor for steel. It neither protects, nor attacks copper and aluminium alloys.

Molybdates (anodic)

a) the molybdates are very rarely used alone, except at very high concentrations. These are anodic inhibitors, when they are used with the zinc at a pH of 7.5;
b) they make part of the products considered to be toxic. Their use is reserved to the treatment of closed circuits or the circuits whose concentration ratio is very high;
c) they are too expensive, sometimes associated with the nitrites at alkaline pH.

Polysilicates (cathodic)

a) the polysilicates and orthosilicates have no effect on the biological treatments of wastewater, and they are not affected by the reducing conditions;
b) their performance is not good at the pH inferior to 7.0, but they are excellent in the water with low concentration of suspended solids, at the values of the pH superior to 7.5 and until 9.0. In general, they are used with other inhibitors.
c) they should not be used, if the concentration of natural silica is superior to 150 ppm;
d) the protective film is difficult to form and easy to destroy;
e) they protect copper and aluminium alloys.

Orthosilicates (anodic)

a) they are not very efficient, even at high concentration;
b) they can contribute to a serious pitting.

Nitrites (anodic)

a) in general, they are not used alone. They can cause an attack by pitting. They necessitate the high pH (superior to 7.5). Nitrite may have been the first product to be used as anodic inhibitor. High concentrations, superior to 500 ppm, are necessary. These are excellent inhibitors for the closed systems, but they are not convenient for the circuits with tower because of their fast oxidation into nitrate and due to the fact that they serve as food for microorganisms (when transformed into nitrates);
b) they are very toxic for animals, as well as for human beings.

Phosphonates (cathodic)

We must note that they have an inhibiting action close to that of polyphosphates, but with a resistance to hydrolysis extremely superior to that of polyphosphates. It is, however, their scale-preventing action that presents by far the main interest. They attack copper.

Certain phosphonocarboxylic molecules present excellent anticorrosion and scale-preventing properties. Their low toxicity constitutes a supplementary advantage.

Modified tannins and lignins (apolar)

Different types of tannins and lignins are used in the anticorrosion treatments; the pattern of their action is quite imprecise: sometimes they avoid the cathodic depolarization. In general, they form a waterproof film or they rather improve the quality of the films. They are used in association. They never provoke pittings, even at high temperature.

Aromatic nitrogens

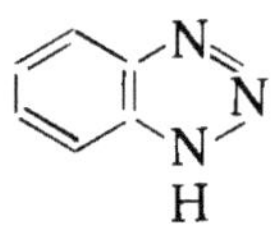

Benzotriazole

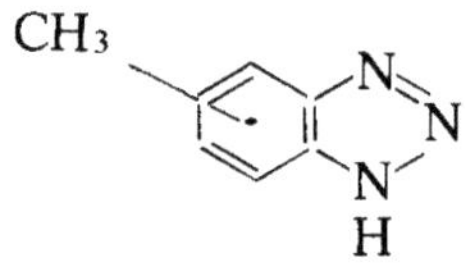

Tolyltriazole

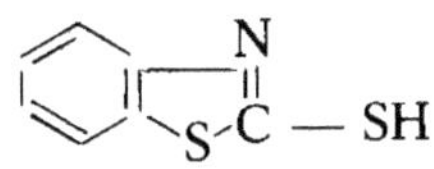

Mercaptobenzothiazole

Mercaptobenzothiazoles

These are corrosion inhibitors mainly for copper and its alloys. They provide an excellent protection even at very low doses.

The mode of their action is the formation of a covalent complex at the surface of the metal, which reinforces the normal protection of copper oxides or carbonates already formed.

Disadvantage of these inhibitors: the sensitivity to oxidizing agents.

Triazoles

These are corrosion inhibitors for copper and its alloys. They are stable at/to the oxidizing agents.

Synergetic mixtures

Most often the mixture of two or more inhibitors is used, which has the effect of extending the limits of use and reducing considerably the total dose required, which is interesting when it is a matter of heavy, toxic metals, whose discharge must be minimized as far as possible. Zinc polyphosphate mixture (cathodic).

The presence of zinc retards the inclusion of iron in the film and makes it more protecting, while remaining thin, which is interesting when high heat transfer is desired. This system functions well in the area of neutrality. It has a dispersing effect due to the polyphosphates.

Commonly used dosages:

10 - 20 ppm in PO_4 and 1 - 3 ppm in zinc

- Zinc phosphonate mixtures (cathodic)

They allow operating at the higher pH, compared to the products discussed above, due to the very significant scale-preventing effect of the phosphonates. In order to avoid the copper corrosion by the phosphonates, we add an inhibitor of the triazole family, when this metal and its alloys are present in the circuit.

- Zinc tannin or zinc lignin mixtures

The cathodic action of the zinc and its fast film formation are added to the advantages of lignosulfonates. It is often used in the multi-metal systems with an organic inhibitor for copper.

Commonly used doses: 3 to 5 ppm in zinc.

- Phosphonate-acrylate or polyphosphate-acrylate mixtures

This system is less subjected to the disadvantages of insufficient dosing for the high pH. It is particularly efficient due to its dispersing scale-preventing effect.

- Zinc-phosphonate-polyacrylate mixtures

These formulas have remarkable anti-corrosive and antideposition properties in the same time.

CORROSION COUPON

I No pH treatment = 9
II Treatment with phosphonate only pH = 9
III Treatment phosphonat/silicate pH = 9
IV Treatment phosphonate/chromate pH = 9
V Treatmenchromat/phosphate/zinc pH = 9

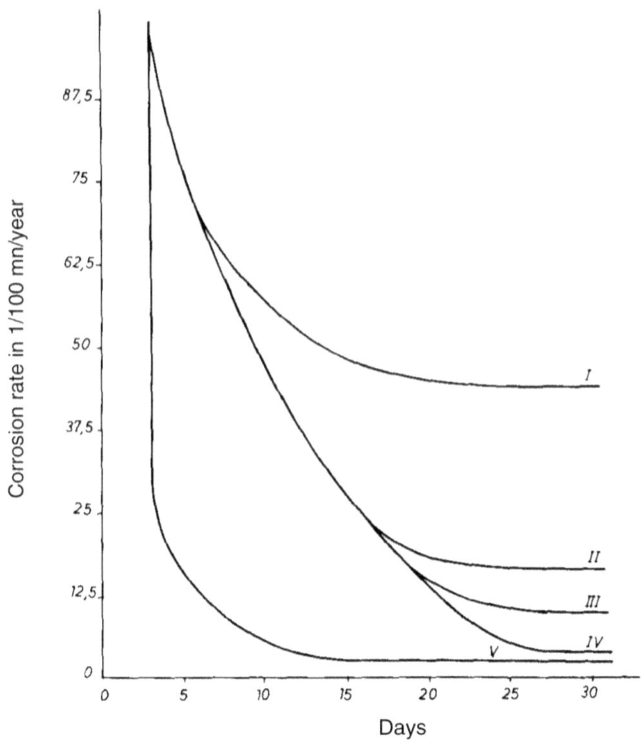

9

Microbiological Fouling and Its Treatment

Microorganisms are present everywhere in nature, be it in the air, in water or in the soil, and they play a fundamental role in the permanent recycling of matters there, which constitutes the life of our planet.

The microbial life is subject of interest for many industrial processes.

It is used, deliberately, in the agri-food industries, the pharmaceutical industries and some chemical industries, for producing the molecules having resisted, until present, to the attempts of chemical synthesis. It has increasingly been used for the water purification, where it consumes the by-products undesirable for our activity.

On the other hand, the microbial life invades the aqueous systems used by the industries; when the conditions for life favorable to one or more species are realized, this invasion is the source of problems, which can be divided into three types:

- fouling of the equipment, which makes the conditions of their operation deteriorate;
- dispersion, through the aerosols generated by cooling towers, of the pathogenic bacteria, Legionella pneumophila;
- corrosion of the elements in contact (exchangers, piping), which shortens their service life.

9.1 Microbiology of the Cooling Towers

It is important to see that, although an isolated cell has no effect whatsoever on the functioning of the cooling towers, the conditions in cooling systems are most often favorable for the development of microorganisms: pH close to neutrality, tropical temperature and the abundance of mineral salts and organic compounds serving as nutrients, while the light being capable to meet needs for the photosynthesis in the towers.

Therefore, even though the water concentration ratio generally does not exceed 6, the microorganism concentration may exceed several million.

The microorganisms found in water cooling circuits belong to three main groups.

9.1.1 Bacteria

These unicellular and microscopic living organisms can not be classed in the plant or animal kingdom, since the differentiation between the two kingdoms requires more advanced state of evolution.

Numerous bacterial species secrete polysaccharide which acts as a binding agent in the deposit formed on the heat exchange surfaces. This is *the biofilm*. The biofilm serves as the hanging support for suspended solids, the colloids present in the water. It is a place of life for all the bacteria species, including the Legionella pneumophila.

The true weight of the bacteria themselves in these deposits is much lower than 1 % of the total weight.

The presence of a large number of bacteria in a cooling circuit may have no effect on the system. Problems occur when debris agglomerates in the sludge and coverings on the heat exchanging surfaces.

There exists a very large number of bacterial species with very different characteristics:

► *Colony formation*

Some of them are unicellular, in other words, the bacteria remain isolated. Certain among them form colonies by remaining in more or less close proximity and surrounded by various secretions which prevent their moving apart.

The size of the bacteria may vary from 0.1 to 80 μm.

► *Form of bacterial cells*

- Spherical cocci;
- Bacilli or rods;
- Spirilla with different degrees of curvature.

► *Mode of reproduction*

- Spore formation (form of resistance);
- No spore formation (encapsulated).

► *Optimum temperature*

- Psychrophilic bacteria	from 0 to 25 °C;
- Mesophilic bacteria	from 25 to 45 °C;
- Thermophilic bacteria	from 45 to 70 °C.

► *The presence of oxygen*

Aerobic bacteria can live only in an aerated medium, as they require air to ensure their metabolism of oxidation-reduction (redox).

Anaerobic bacteria can not live in air and use other chemical processes for their oxidation-reduction metabolism. (indifferent bacteria).

► *Carbon source*

Bacteria may be autotrophic, that is to say, they do not need organic matter, as they utilize carbon dioxide as carbon source.

They may be heterotrophic, using both inorganic and organic carbon.

They may be parasitic, using living matter, or saprophytic, living on dead organic matter.

In order to understand their harmful effects in cooling systems, a certain number of interesting points should be noted.

Encapsulated aerobic bacteria, such as Aerobacter, Pseudomonas, Flavobacterium are generally associated with problems of deposits, as they are surrounded by a polysaccharide mucus, which protects them from the environment and serves as food in case of food shortage.

Aerobic, spore forming bacteria, such as Bacillus subtilis ou Bacillus cereus are often less fouling, but their capacity to form spores makes them more resistant to biocides.

As for the sulfur bacteria, they are related to corrosion problems as well.

Thiobacillus thiooxydans (aerobic) oxidize sulfur, sulphides or sulphates, and may cause the pH of the medium drop to 1. Moreover, they live symbiotically with sub-adjacent anaerobic bacteria.

Desulfovibrio desulfuricans, which are sulphate-reducing bacteria, transform sulphates into sulphides, according to the following reactions:

$$10\ H^{+} + SO_4^{2-} + 4\ Fe \rightarrow H_2S \uparrow \ + 4\ Fe^{++} + 4\ H_2O$$
$$\text{et } H_2S + Fe^{2+} \rightarrow FeS \downarrow + H_2$$

Given the fact that sulphates are present in almost all waters, there is a constant risk of proliferation of this species.

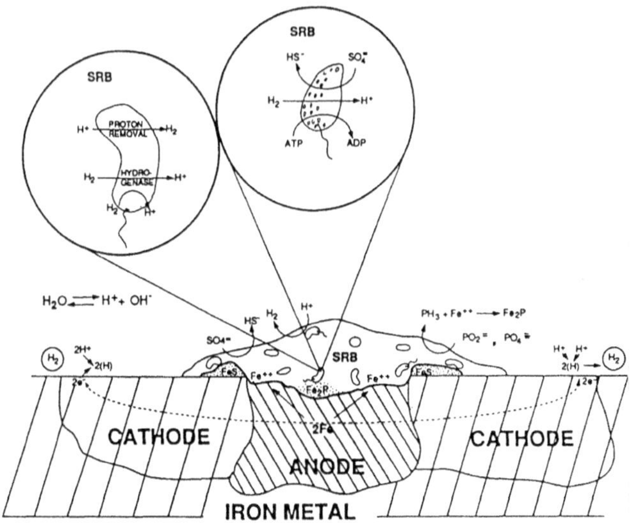

Mechanisms of corrosion by sulphate-reducing bacteria

Iron bacteria: Gallionella, Crenothrix, etc. utilize iron as an energy source for their cell metabolism according to the following reaction:

$$4\ Fe\ CO_3 + O2 + 6\ H_2O \rightarrow 4\ Fe\ (OH)_3 + 4\ CO_2 + 81{,}000\ cal$$

Enormous quantities of iron hydroxide form, causing considerable fouling. These bacteria may favor the formation of pitting and tuberculation on the equipement made of carbon steel.

The bacteria

Type	Species	Physical characteristics	Problems caused	Required pH	Required temperature
Sludge formers :					
Do not form spores Aerobic, Encapsulated	Pseudomonas Flavobacteriu Aerobacter Entobacter Mucoides Klebsiella Proteus	Gelatinous Flocculant Mucilaginous Gray-yellow	Fouling Protect corrosive bacteria Insulate from inhibitor	4 to 8 optimum:7.4	20 to 40 °C
Aerobic Spore formers	Bacillus subtilis Bacillus	Gelatinous May be fibrous, filamentous, yellow, gray or orange	Fouling Protect corrosive bacteria More difficult to kill	5 to 8	20 to 40 °C
Corrosive : Depositing iron	Gallionella Crenothrix Leptothrix Sphaerotilus Streptomyces	Red-brown Sludge Filamentous Aerobic	Fouling Doposit iron oxide Increase corrosion	7.4 to 9.5	20 to 40 °C, even 10 to 60 °C

The bacteria (*continued*)

Type	Species	Physical characteristics	Problems caused	Required pH	Required temperature
Sulfuric acid producers (aerobic)	Thiobacillus thiooxydans	Aerobic	Metal corrosion	1 to 6	20 to 40 °C
Nitric acid producers	Nitrobacter Nitrosomonas	Aerobic	Metal corrosion		20 to 40 °C
Sulphide producers (sulphate reducers) Aerobic	Desulfovibrio Clostridium Thiobacillus	Black sludge Forms spores Anaerobic .	Metal corrosion Reduce chromate Precipitate zinc Form gases May cause repugnant odors	4 to 8	20 to 40 °C

The nitrifying bacteria
These bacterial species can transform ammonia and ammonia derivatives into nitric acid. Due to the low pH produced, they attack most metals, especially mild steel. They are difficult to analyse, since they coexist with all kinds of microorganisms.

The most often found are: Nitrobacter and Nitrosomonas.

Bacteria of the type Legionella
A special chapter is dedicated to this bacteria family because of serious health problems that their presence in cooling circuits with tower may cause.

9.1.2 Algae

These are lower plants characterized by the absence of sap conducting tissues as found in the higher plants, but they all contain chlorophyll, as opposed to fungi. They may be unicellular or multicellular and exhibit a large variety of forms.

They are all characterized by their needs for water, air and light for their development. Thus, they are found in open basins, especially on the surface, and on the structures of cooling towers.

In terms of form, they may be unicellular or multicellular, often filamentous:

- *the filamentous algae* such as *Oscillatoria* may reach 1 metre in length;
- *the diatoms* are unicellular and use the silica in the water to construct their outer skeleton;
- *the blue-green algae,* such as *Calothrix* can fix ambient oxygen, transforming it into nitrate nitrogen.

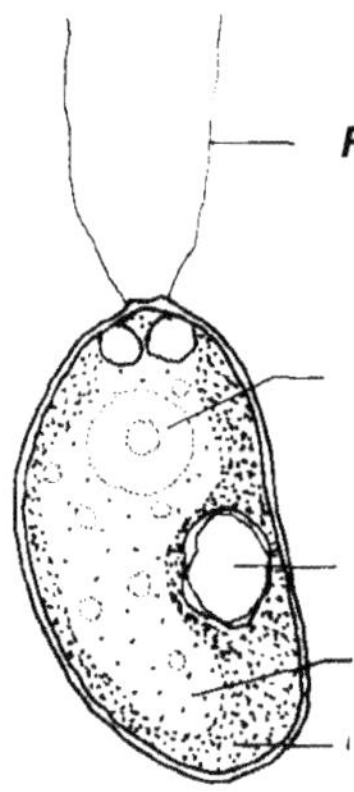

CELLULAR STRUCTURE OF AN ALGAE IN DIAGRAM

In all cases, they produce oxygen as all other plants containing chlorophyll, thus contributing to accelerated corrosion, in addition to the problems that they cause through fouling.

The algae

Type	Species	Physical characteristics	Problem	Required pH	Required temperature
Green algae	*Chloroccus* *Ulathrix* *Spirogra* *Synedra* *Scenedesmus*	Organisms of large size, rubbery, viscous, soft structure.	Massive vegetation interferes with appropriate water distribution.	5.5 to 8.9	30 to 35 °C

The algae (*continued*)

Type	Species	Physical characteristics	Problem	Required pH	Required temperature
		Fabricate their own alimentation. Green to blue-green color. Require light to survive.	Fouling. Trap suspended matters. Protect Corrosive bacteria.		
Blue-green algae	*Chlorococcus* *Oscillatoria* *Anacystis* *Stichococcus* *Phormidium*			6 to 8.9	35 to 40 °C
Diatoms	*Cyclotella* *Navicula* *Flagilaria* *Diatoma* *Asterionella*	Utilize soluble silica as source for exoskeleton		5.5 to 8.9	18 to 35°C

9.1.3 Fungi

These are plants which lack chlorophyll, but they, as algae, have a rigid cellulose membrane. They may be constituted by isolated cells, colonies of isolated individuals or filaments with varying degrees of ramification.

In nutritional terms, their lack of chlorophyll makes them dependent on other organisms which can provide them with an organic carbon source. In particular, they decompose the carbonaceous matter of wood, thus contributing to the maximal destruction of the material from water treatment installations.

Their reproduction may be asexual by simple vegetative multiplication of cells, or may be sexual with spore formation. These spores are forms resistant to the unfavorable environmental conditions and may remain inactive during long periods until conditions again become favorable. Moreover, they are very resistant to biocides, but in the case where biocides force the fungi to remain in this form, there is no fungal development and this may be sufficient.

The fungi

Type	Species	Physical characteristics	Problems	Required pH	Required temperature
Fungi (filamentouses)	*Aspergillus* *Alternaria* *Penicillium* *Fusarium* *Mucor* *Streptomyces* *Cladosporium*	Filamentous, furry or enmeshed White or gray. Form spores in most cases.	Superficial rotting of wood. Fouling. Form corrosion cells. Contribute to sludge.	2 to 8 optimum 5 to 6	0 to 40 °C
Yeasts	*Trichoderma* *Torula* *Monilla* *Verticillum* *Saccharomyce* *Endomyces* *Oospora*	Leather-like skin or rubbery (in certain cases). May form protective layers resembling spores.	Contribute to sludge. Fouling. Form corrosion cells.	2 to 8 5 to 6	0 to 40 °C
Basidiomycetes (higher fungi)	*Poria* *Lenzites*	Flat surface fungi. White to Brown May form spores	Serious internal rotting of wood.	2 to 8 optimum 5 to 6	0 to 40 °C

The proliferation of microorganisms in cooling circuits very often leads to the formation of a biofilm which adheres to the surface of pipes and various equipments constituting the system. The biofilm causes considerable problems, which we will discuss below.

9.1.4 Biofilm

As we have seen, it is a matter of biological deposit, which adheres to the solid surfaces:

- walls of the equipment, pipes and zones where the water concentration is varied;
- suspended particles.

These biofilms are known, as they cause bacterial contaminations of the circulating water through liberation of the adhesive bacteria.

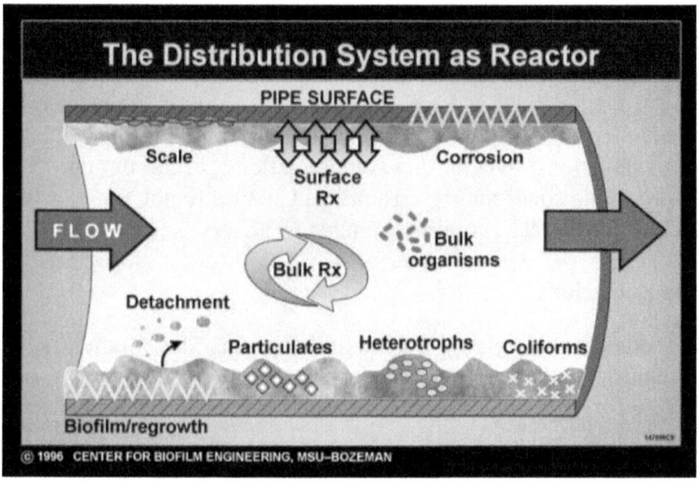

Biofilm in the circuits (doc. CBE, University of Montana)

Among the microorganisms liberated, certain, such as Legionella may undermine health of the persons in contact with the cooling systems.

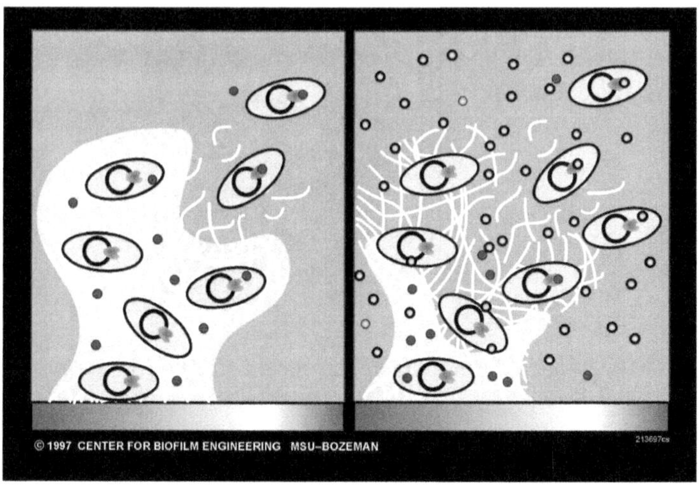

Liberation of the biofilm (doc. CBE, University of Montana)

The biofilm control is difficult, as it provides a resistance to the penetration of the active matter of biocide used. In terms of prevention, it is proved to be better to use a combination of a dispersing agent and a biocide. Both of them may be present in the same formulation. In case of the treatment based on oxidizing biocide, the dispersing agent is added separately either on continuous or discontinuous basis.

9.2 Legionella Pneumophila Bacterium

The legionnaires' disease was described first in 1976, following an epidemic outbreak which had affected more than 200 participants of the 58th convention of the American Legion in Philadelphia, and caused death of 34 persons. Since then, several epidemics have been observed in collective buildings, hotels, hospitals and industrial installations, where the contamination occurs through atmospheric cooling towers, air conditioning systems and domestic hot water networks. The declaration of this disease is obligatory since 1987.

9.2.1 The Bacterium

The Legionella are bacteria of hydrotelluric origin. This Gram negative bacilli familly includes 43 species and 64 different groups. *Legionella pneumophila* causes 90 % of legionellosis and the serogroup 1 of this species (Lp1) is associated to more than 80 % of all cases.

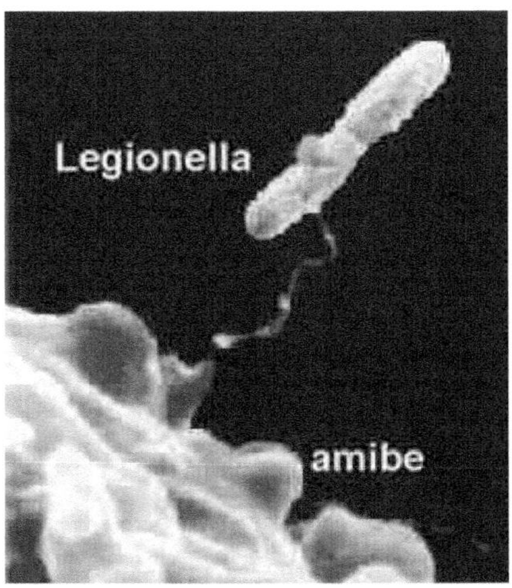

Legionella developing on an amoeba (IRH document)

The Legionella proliferate when the water temperature is between 25 and 45 °C. They survive below 25°C and multiply beyond this temperature. Around 50°C, destruction occurs (in some hours). Near 60°C the time of destruction becomes very short (of the order of some minutes). The diagram below shows the mortality (destruction) rate of *legionella pneumophila* depending on temperature. The bacteria tolerate a wide range of pH (6 to 10). They proliferate in the presence of high concentrations of calcium, magnesium, metallic residues and certain materials such as rubber, polyvinyl chloride, polyethylene or silicone, as well as other microorganisms of aquatic environment such as cyanobacteria or free amoeba. They find refuge in the *biofilm* that we have studied in the previous chapter.

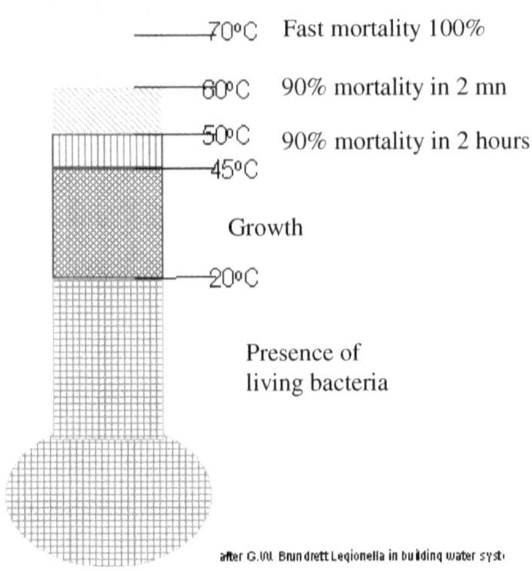

Mortality rate of the *legionella pneumophila* depending on temperature (according to G.W. Brundrett)

Result of the bacteria detection is expressed in colony forming units (CFU) per litre. The current limit of detection is from 50 to 100 CFU/l depending on laboratory. The bacterial growth in the cultivation medium is relatively slow, varying from 5 to 10 days depending on species.

The PCR (polymerase chain reaction) method presents the advantage of providing information on the number of Legionella present 3 hours after the analysis realization.

It will probably allow legionella risk control thanks to a fast result. The short delay for response between the sampling and the result of the analysis can allow adjusting the treatment before the degree of legionella concentration becomes dangerous for humans.

As the legionella are present in all water medium, it is difficult to prove the contamination. For this purpose, it is necessary to isolate the strain from the patient (which is made in 20 to 30 % of all cases) and compare it, using the methods of genetic analysis, with various environmental strains sampled. We can say that the bacteria of this family are found all around us in daily life.

9.2.2 Contamination and Pathology

The contamination of persons exposed to the bacterium occurs mainly through the inhalation of tiny water droplets (size inferior to 5μm) which are contaminated and diffused by aerosol. These aerosols reach the pulmonary alveoli, infest the pulmonary macrophages and provoke their destruction. The legionellosis is not a contagious disease.

The legionellosis takes two distinct clinical forms:

- **Pontiac fever** is a minor form passing most often unnoticed.
- **Legionnaires' disease** appears in the form of a serious pulmonary infection (mortality of 15 to 40 %), after a minimum of 2 to 5 days of incubation.

Risk to develop legionellosis is quite variable and depends on many individual factors, permanent or temporary, such as the age, alcoholism, nicotinism, immunodeficiency and chronic respiratory infections. The risk is more significant in men than in women.

9.2.3 Installations at Risk

The Legionella are present at natural state in fresh water (lakes and rivers) and in humid soils. From natural environment, the bacterium colonizes artificial water sites when the conditions for its development are created. Its proliferation may be favored by the conditions present in so-called "installations at risk" such as hot water network, cooling circuits with atmospheric cooling towers, whirlpool bath (spa), humidifiers for air conditioning systems, etc.

Studies of different epidemics and isolated cases allowed associate the legionnaires' disease to two main sources of contamination: cooling towers and domestic hot water distribution systems.

Domestic hot water systems: contamination through an aerosol generated by the use of domestic hot water (mainly when taking shower) is at the origin of a large number of legionellosis cases.

The Legionella risk control in domestic hot water networks is carried out through an audit of installations, which results in an adjustment of the system concerned (elimination of oxbow), allowing a nearly constant circulation in order to limit the zones of biofilm formation (temperature below 60°C), as well as a preventive oxidizing treatment compatible with the materials used, the durability of the installation and the use of water for sanitary pupose.

9.2.4 Atmospheric Cooling Towers

Cooling water circuits with atmospheric cooling tower, used for air conditioning systems and in industries, create conditions favorable to the development of Legionella, due to the water temperature and the contact with air. The bacteria may develop in the circuit water, then it may be transferred into the outer air through the windage loss in the form of microdroplets. Deposits due to scale and corrosion products, as well as biofilms, which are very often present on the surfaces, favor to great extent the development of these bacteria. We have seen in the chapter 6 the description and functioning of the atmospheric cooling towers.

The formation of aerosols causing the Legionella contamination has 2 origins:

- the windage loss;
- water carry-over in the windy zones.

The windage loss concerns only so-called "humid" cooling towers (energy evacuation through water evaporation).

The windage loss exists in different types of cooling towers that we have previously described. Studies concerning this phenomenon have shown that:

— the gas ejection velocity, which varies between 3 and 10 m/s has effects on the plume dispersion (the tower height as well);
— the percentage of windage loss varies between 0.1% and 0.0005% of the recirculation water flow.

Parameters which explain this difference are mainly the following:

— the quality and the degree of cleanness of the droplet separator;
— the increase of recirculation flow in relation to the nominal flow initially considered.

The following table shows the impact of the windage loss on the Legionella propagation. The calculations were made for a tower having a recirculation flow of 1,000 m^3/h.

Concentration of Legionella (CFU /l)	Windage loss rate (%)	Quantity of aerosols/h (kg/h)	Quantity of Legionella discharge
500,000	0.0005	5	250,000
500,000	0.1	1,000	500,000,000
1,000	0.0005	5	5,000
1,000	0.1	1,000	1000,000

Table. Windage loss impact on the Legionella discharge (doc. IRH)

The information provided by this table shows that a well equipped tower, even highly contaminated, will discharge less Legionella into the atmosphere in the form of aerosols than a badly designed or fouled tower which cools water containing 1,000 Legionella per litre. We also see that the degree of Legionella concentration in the circulation water (which respects the requirement imposed by the legislation, 1,000 CFU/l) is not sufficient for evaluating the health hazard.

9.2.5 Risk Analysis

This analysis allows defining an action plan aiming to reduce the risks of Legionella contamination. The analysis concerns the design, the maintenance and monitoring of the systems. Its implementation makes a part of the elements imposed by the order of 13th December, 2004.

► *The methodology*

a) Audit of the installation design
- Circuit characterization:
  - Nature of the towers
  - Number of towers
  - Materials present in the circuit
  - Nature of the make-up water, the concentration rate
  - Water circulation system
  - Basins, oxbows, low points.
- Characterization of the towers
  - establishment
  - power
  - number of fans
  - nature of the exchanger
  - presence of droplet separators, inspection manhole

b) Audit of the maintenance procedures and monitoring book
- Frequency and nature of the operations of cleaning, draining, descaling and disinfection of the circuits and towers;
- Analysis of the analytical physico-chemical and microbiological monitoring realized on the circuit.

c) Analysis of the water treatment applied

Nature of the scale-preventive, anti-corrosion and biocide treatments

d) Proposition of the action plan

Depending on the audit results, the critical points are identified to propose appropriate actions concerning the following:

- Maintenance operations ;
- Evaluation of the efficiency of operations and planning of the relevant measures;
- Plan of the physico-chemical and microbiological monitoring;
- Action plan: warning level and action level, as well as the nature of the operations associated.

9.2.6 Treatments Recommended

Treatments with bactericidal action are necessary in these cooling circuits. They are generally associated with scale-preventive and anticorrosion treatments. The most efficient and most frequently-used treatments (both in preventive and curative) are the following:

— chlorination and other chlorinated products;
— bromation;
— organic biocides.

They can be either associated or not with biodispersants.

These treatments are studied in the chapter 9.5.

The heat treatment, through the temperature increase up to more than 60°C in the installations, is not applicable to the industrial systems. It is reserved to the domestic hot water circuits.

Problems encountered in cooling circuits due to corrosion, scale-formation, fouling and microorganisms are associated between them. It is thus necessary to make sure that these phenomena are well controlled.

9.2.7 Monitoring

It includes the following:

- results of all physico-chemical and biological analysis, as well as the corrosion results, the value of the Ryznar index;
- observations made during visits;
- Legionella counting;
- shutdowns of the system operation;
- procedures of cleanings carried out.

To conclude, the Legionella risk control passes through a global approach of the question, including monitoring, management and rigorous maintenance. The design and the establishment of atmospheric cooling towers, the choice of the materials used in the circuit and the conditions of operation should be studied beforehand. An appropriate water treatment should be implemented and it should be the subject of a regular and rigorous monitoring. An evaluation of the efficiency of water treatment processes should also be carried out.

9.3 Control of Microorganisms: Bacteriological Analysis

As we have seen previously, the proliferation of microorganisms in cooling circuits may cause very significant damages to different parts constituting these systems.

It is thus very important to control the level of bacterial contamination in the circulating water in order to better identify the potential problems that may result from it and apply an efficient biocide treatment in order to prevent them.

It is not necessary to proceed to complete identification of the micribial flora and fauna present. In most cases, the counting of total germs, as well as the semi-quantitative research of sulphate-reducing bacteria are sufficient. These analyses allow also determining the best bactericidal treatment to be applied and monitoring its efficiency.

Kits for bacteriological analyses easy to use and capable to determine with a sufficient precision the bacterial activity in the circuit, are commercially available.

Following figure shows the reference tables of this type of analyses (single-use).

The counting of total germs is a good indicator for the degree of cleanness of the circuit.

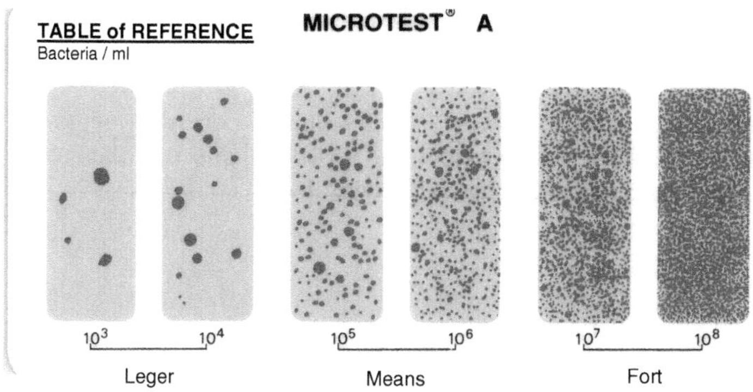

Determination of total aerobic bacteria (doc. France Organique)

Results of the analyses are given in CFU/ml. Following values are generally accepted for evaluating the degree of cleanness and the efficiency of the biocide treatment.

CFU/ml	state of the circuit	treatment
$< 10^3$	very clean	efficient
$10^3 - 10^5$	clean	efficient
$> 10^5$	dirty	necessary
$> 10^7$	very dirty, fouling risk	necessary

The research and counting of the bacteria of the type Legionella have been discussed in the chapter dedicated to them.

Even if there is no direct link between the total germs' concentration and the Legionella concentration, the maximum threshold for the Legionella concentration of 1,000 CFU /l, makes that the concentration level of total population must be inferior to 1,000 CFU / ml.

Following figure shows the reference table related to the research of anaerobic sulphate-reducing bacteria.

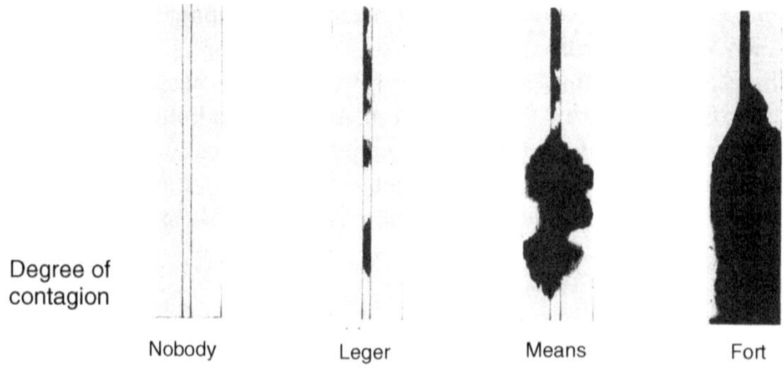

Determination of sulphate-reducing bacteria (doc. France Organique)

9.4 Problems Caused by Biological Growth

9.4.1 Corrosion

Certain migroorganisms found in cooling systems use hydrogen in their metabolic process, causing a cathodic depolarization. Moreover, the oxygen produced by algae also causes a cathodic depolarization.

Although these polarization phenomena are important, the most significant contribution of living organisms to the corrosion process is the sub-jacent erosion they cause. The organic mass which is very often mixed with a large quantity of mineral deposits that it hinders to circulate causes a differential oxygenation of points more or less close to the wall. The masked metallic parts become strongly anodic and undergo intense corrosion. In addition, these deposits locally hinder the action of anticorrosion treatments, which is an additional factor favoring corrosion by pitting.

Many microbiological species exhibit direct corrosive effects themselves. We have seen the sulphate-reducing bacteria, which dangerously release hydrogen sulphide becoming highly corrosive by decreasing the pH. The bacteria which transform sulphur and sulphur derivatives into sulphuric acid have the same effect. Similarly, the nitrifying bacteria decrease pH while producing nitrates, and this is at the expense of inhibitor nitrites in the system.

9.4.2 Deposits

Mucus producing microorganisms, particularly encapsulated bacteria, have a disastrous effect on heat transfer. This highly sticky organic matter forms adhesive deposits, including all mineral deposits which are normally not very scale-forming: silts, oxides, etc.

These deposits are very voluminous, as the analysis shows it. For example, in a deposit containing 20 % organic and 80 % inorganic matter, the organic matter may represent 90 % of the volume of the deposit, because it is constituted by mucus containing 85 to 90 % water.

Similarly, ferric hydroxide producing bacteria are capable of producing several times their volume of hydroxide.

All these accumulations decrease the flow rates and heat exchange capacities, as for the exchange part of the installation. In the evaporation part, that is to say the cooling tower itself, the situation is just as dramatic: water-air contact is seriously reduced, since the water, instead of being divided into droplets, flows on the gelatinous mass, thus reducing heat transfer at this side as well.

9.4.3 Deterioration of Wood

Wood is mainly attacked by different fungal species. Submerged zones are subjected to attack by the less differentiated fungi species, while the exposed parts are attacked by the more evolved species.

The use of wood has progressively been abandoned in favor of the less fragile materials, in particular, plastic materials.

9.5 Biocide Treatments

There are three main types of biocide treatments:

I – Oxidizing type

- Chlorine
- Chlorine dioxide (ClO_2)
- Sodium hypochlorite or Javel water (bleach)
- Bromation
- Others

II - Non-oxidizing type

III - Dispersing agents – Surfactants

9.5.1 Oxidizing Biocides

9.5.1.1 Chlorine Treatment

The first time that chlorine was used for disinfection was in 1894, in a drinking water system. We mention its use in recirculating cooling water systems in 1938, but its use became generalized only a decade later. Nowadays, its use is much decreased.

Gaseous chlorine is totally soluble in water. The relevant reaction is as follows:

$$H_2O + Cl_2 \rightarrow HClO + HCl$$
$$\text{et } HClO \rightarrow H^+ + ClO^-$$

If the pH is lower than 5.0, chlorine is present in the form of chlorine molecules. If the pH is between 5.0 and 6.0, it is present almost entirely in the form of hypochlorous acid, HC1O. When the pH is higher than 6, the hypochlorite ions ClO^- begin to appear, and their concentration increases until becoming predominant, when the pH exceeds 7.5. This is represented graphically in following figure.

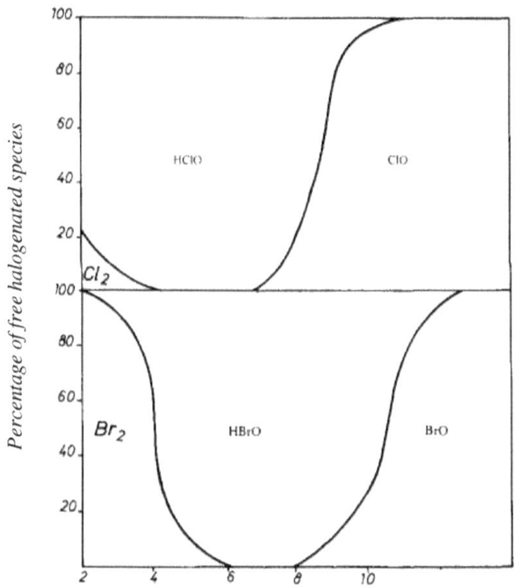

Chemical species encountered in oxidizing halogen solutions at 20 - 25 °C.

It is due to the hypochlorite ion that the chlorine solutions have disinfecting properties. The oxidizing power of hypochlorous acid is very high. However, it decreases with increasing pH.

In addition to the fact that it reacts with water itself, chlorine also reacts with numerous impurities found in cooling water systems.

An important reaction occurs with ammonia and ammonium compounds. Another important reaction occurs with organic matters, yielding toxic chlorinated organic products. Reactions occuring with ammonia are the following:

$NH_3 + Cl_2 \quad \rightarrow$
$NH_2Cl + Cl_2 \quad \rightarrow$

$NHCl_2 + Cl_2 \quad \rightarrow$
$NH_2Cl + HCl$
$NH\ Cl_2 + HCl$
$NCl_3 + HCl$

(monochloramine) (dichloramine)

(nitrogen trichloride)

The course of these reactions largely depends on the pH of the water. At pH higher than 8.5, monochloramine is the only product that forms. When the pH is between 8.5 and 4.5, mixtures of monochloramine and dichloramine form. At the pH values lower than 4.5, the prevailing product is nitrogen trichloride.

Chlorine dissolved in water has powerful oxidizing properties. Besides ammonia, it can react with other organic and inorganic compounds, for example, oxidizable metals such as iron and manganese salts (naturally present, added for clarification or issued from the corrosion of steel components of the system). Nitrites are oxidized into nitrates.

There are numerous organic substances present in cooling water, which react with chlorine. They may be issued from wood, may be pollutants of the raw water such as organic iron, lignins and humus, may be pollutants from the system, such as phenols, light and heavy hydrocarbons and various chemicals, without counting the chemicals used for the water treatment.

These organic substances form chlorinated organic substances, some of which resulting from the addition or substitution of chlorine, are toxic and have bactericidal, fungicidal or algicidal properties. These toxic compounds may constitute a considerable advantage for the chlorine treatment program, particularly at high pH values (from 7.5 to 9.0).

Since chlorine gas dissolved in water forms acids (hypochlorous and hydrochloric), these acids react with the alkalinity of the water and thus tend to decrease the pH. Each part of chlorine added reduces alkalinity by about 1.22 ppm the alkalinity (in the form of $CaCO_3$). If the chlorine demand becomes relatively high (20 to 30 ppm), the resulting low pH values (lower than 5) may significantly contribute to corrosion of the metallic component. Addition of alkaline agents may then be necessary in order to maintain the pH near neutrality.

These reactions are the basis of what we call the point of rupture of the chlorine treatment (chlorination). When the dose of chlorine is increased, the chlorine residue apparently increases also, since most tests can not distinguish between chlorine and hypochlorous acid, as well as chlorine and chloramine. During this time, ammonia nitrogen remains constant. However, when the chlorine residue reaches a certain threshold (about 8 times the ammonia nitrogen concentration), it drops rapidly, as does the ammonia nitrogen. This is due to the formation of nitrogen trichloride.

The *relative efficiency* of chlorinated compounds as oxidizing biocides is as follows:

$HClO$: ClO : NH_2Cl : $NHCl_2$ correspond to about 1: 2 : 6 :12, *for identical times of contact.*

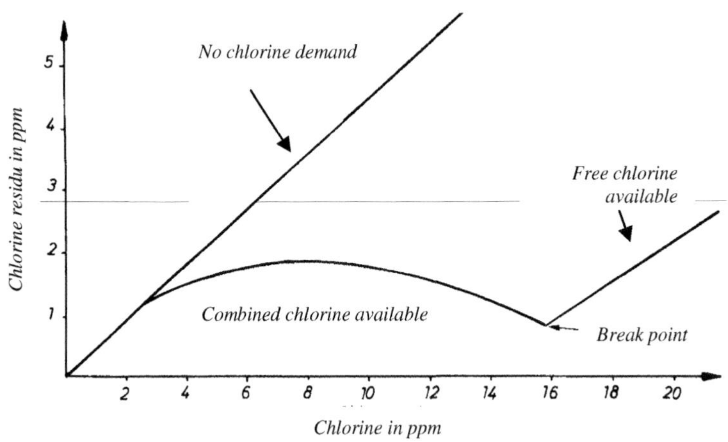

9.5.1.2 Chlorine Dioxide

Chlorine dioxide is a reddish yellow gas formed by the reaction between a chlorine solution or hydrochloric acid and a sodium chlorite solution at pH at most equal to 4.0.

Given the fact that chlorine dioxide is instable, it is more practical to produce it in the vicinity of its utilization.

A sodium chlorite solution is mixed with the chlorine water provided by a chlorometer or with an acidified calcium solution or with sodium hypochlorite in a special generator.

$$Cl_2 + 2\ NaClO_2 \rightarrow 2\ ClO_2 + 2\ NaCl$$
$$5\ NaClO_2 + 4\ HCl \rightarrow 4\ ClO_2 + 5\ NaCl + 2\ H_2O$$

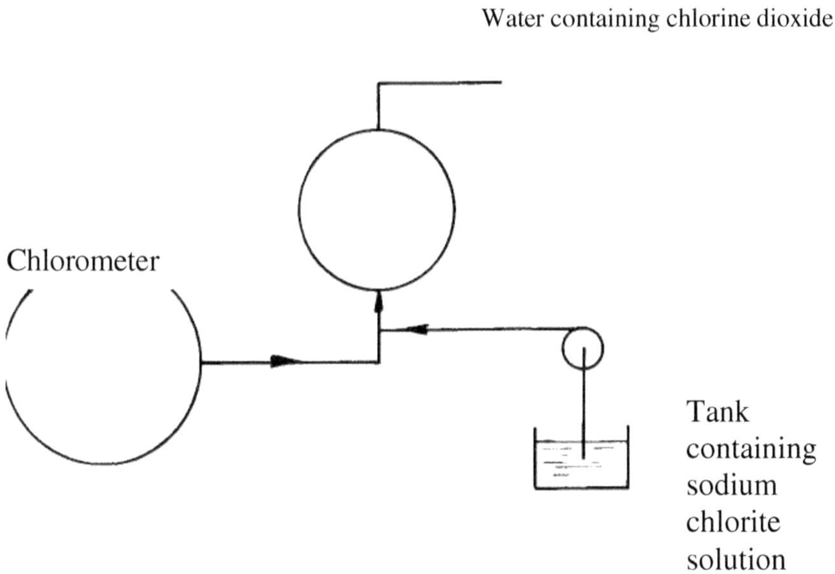

Production of chlorine dioxide

The theoretical (weight) ratio between chlorine and sodium hypochlorite is about 1/3. However in practice, this ratio must be lower than 1/2.

Chlorine dioxide has two advantages, as a biocide, in comparison to chlorine: it does not react with ammonia and its biocidal activity is not much affected by pH variations between 6.0 and 10.0. It has slightly better algicidal properties than chlorine. Chlorophyll in the algae is destroyed by chlorine dioxide, and the water escapes from the protoplasm, thus the plant dies.

However, it is more costly than chlorine.

9.5.1.3 Bromation

Bromation is a very interesting alternative technique in relation to chlorination, and its use is increasing. Different precursors of bromine can also be used:

- gaseous bromine;
- bromine chloride;
- bromochlorodimethylhydantoin (BCDMH);
- bromine salts, and more particularly, sodium bromide associated with sodium hypochlorite (Javel water). The latter is very widely used in cooling circuits, which operates at free pH.

Whatever was the precursor used, in all cases, the active molecule is in fact the hypobromous acid formed by the following reactions:

brome: $Br_2 + H_2O \rightarrow HBrO + HBr$
sodium bromide: $NaBr + HClO \rightarrow HBrO + NaCl$ et
$HBrO \rightarrow H^+ + BrO^-$

The hypobromous acid is dissociated into the H^+ ion and hypobromite ion. HBrO and BrO^- are in equilibrium and their proportion depends on the pH.

In the cooling water systems maintained at alkaline pH, the hypobromous acid is more active, because it is more stable than hypochlorous acid, as the following table shows it:

Percentage of HClO and HBrO (free) depending on pH

pH	%HClO	%HBrO
7.0	90	100
7.5	50	95
8.0	25	80
8.5	10	60
9.0	5	35
9.5	0	10

Bromation has a certain number of advantages compared to chlorination. During nitrogenous pollutions, although the reaction of hypobromous acid with nitrogenous or ammonia derivatives is identical to that of chlorine, it has been shown that this reaction is reversible, as the bromamines formed are not persistent compounds.

The action of bromine is faster than that of chlorine. It is less volatile than chlorine and it is maintained longer and more easily in solution.

Bromation is more efficient at high pH values. In fact, the proportion between $HBrO/BrO^-$ and $HClO/ClO^-$ is totally different depending on the pH, as the above table indicates it.

Hypobromous acid is also much less corrosive than its chlorinated equivalent. It is less harmful for environment.

These advantages largely compensate for the considerably higher cost compared to the chlorination treatment (between 1.5 and 2 times). In addition, it is necessary to have a chlorinated source in order to generate hypobromous acid.

In practice, bromation is realized by a mixture of hypochlorous acid and hypobromous acid generated by the reaction between excess Javel water and sodium bromide. The ratio HClO/NaBr is determined, and this ratio varies between 1 and 5 to 10. Knowledge of the value of this ratio is fundamental. In fact, the more the ratio is close to one, the higher will be the concentration of hypobromous acid and the more significant biocide effect.

9.5.1.4 Other Oxidizing Treatments: Ozone

Ozonation has been used in Europe for water disinfection, since 1902. It is only recently that it has been used in the United States.

Ozone, an allotropic form of oxygen, is the second most powerful oxidizing agents that we know. It is a molecule consisting of three oxygen atoms, whose symbol is O_3. In the usual conditions of use, it is in gaseous state and soluble in water. The oxidation potentials of some biocides are shown below:

Ozone	2.07	Chlorine	1.36
Hydrogen peroxide	1.77	Bromine	1.07
Hypobromite	1.59	Chloramine	0.75
Hypochlorite	1.49	Iodine	0.50
Hypoiodite	1.45	Oxygen	0.40

Due to its low molecular weight, one kilogram of ozone contains more molecules than a kilogram of any other biocide. The following list shows the number of molecules (x 10^{18}) present in one litre of a 1 ppm solution of the main biocides.

Ozone	12.6	Hypobromite	6.2
Chloramine	11.7	Hypoiodite	4.2
Hypochlorite	11.5	Bromine	3.8
Chlorine	8.5	Iodine	2.4

The ozone manufacturing process consists in passing oxygen or air between two electrodes between which a high voltage AC potential difference is maintained. The two electrodes are supplied by a step-up transformer.

Often, one of the electrodes is grounded. A dielectric screen prevents sparkling and maintains the discharge as a regular crown. The latter is visualized as a violet-blue glow. When the oxygen molecules pass into the crown-shaped flux, they become charged and dissociated into ions. These ions recombine into molecules, some of which contain more atoms (ozone) than the two initial atoms (oxygen).

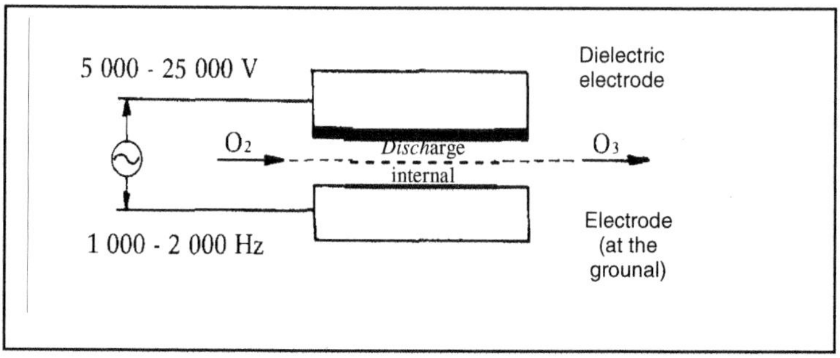

An ozone unit schematically includes the following elements:

- air ventilator and filter;
- a cooling circuit which absorbs the excess heat produced by the discharge;
- an air dryer allowing obtain a dew point of -50°C to -70°C;
- an azone generator;
- an ozone injection system.

Ozone dissolving in water is a key element of a good installation. Indeed, temperature influences much the gas dissolving rate: The ozone solutions are subject to the Henry law. The quantities dissolved are function of temperature and pressure. Moreover, the more significant the gas/water interface, the better is the dissolving.

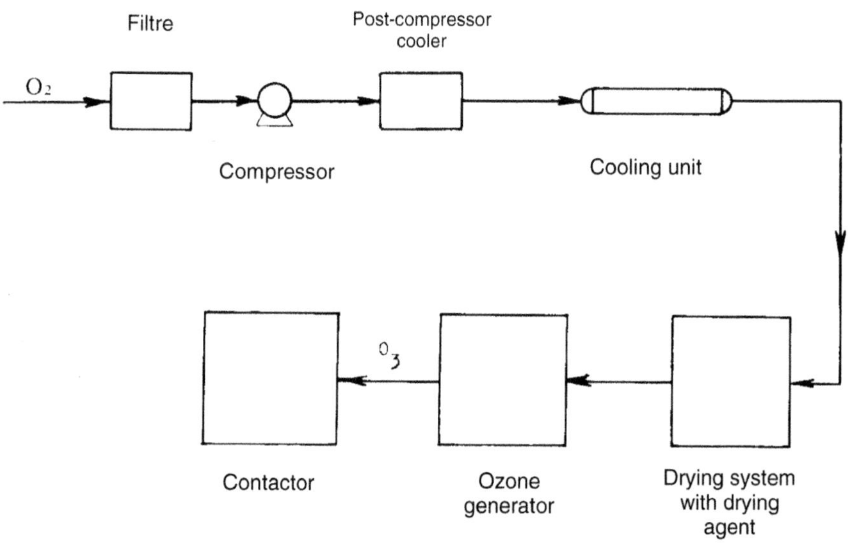

If we use a current of ambient air, about 1 % of the oxygen is converted to ozone, and, if pure oxygen is used, 1.5 to 2 % of it is converted to ozone. When air is used as oxygen source, the energy requirements for the process are 10 - 13 kWh per kg of oxygen transformed.

Ozone is very destructive for vegetation. The ozone residue in a working area must not exceed 0.1 mg/l.

Ozone is a better biocide than chlorine for the following reasons:

— ozone does not react with ammonia;
— ozone is not affected by pH, although a pH of 6.0 to 7.0 is necessary for optimum ozonization;
— ozone requires 5 minutes of contact, while chlorine requires 30 minutes;
— when chlorine is used, living organisms disappear progressively as the chlorine dose increases. Ozone has a flash point threshold at the concentration of 0.4 - 0.5 ppm, above which no living organisms subsist.

It is possible to use coupled techniques (ozone - UV) which have a better efficiency.

Ozone unit (doc. Ozomax)

9.5.2 Organic Biocides (Non-surfactant, Non-oxidizing)

In many cases, the non-oxidizing biocides seem to be more effective than the oxidizing agents. They are often used at the same time or, even better, alternatively with oxidizing agents so as to ensure a broader spectrum of action. If the biocides are sensitive to oxidizing agents, we use an alternate treatment; if they are insensitive to those agents, a minimal dose of oxidant is maintained, with a more massive addition of organic biocides at regular intervals (several days).

Numerous molecules have been abandoned during the latest years by reason of their very toxic character and their negative impact on environment. To remind, we mention the following:

- *chlorinated phenols,* such as pentachlorophenol;
- *methylene bis-thiocyanate;*

- *sulfones* and *thiones;*
- *organo-tin compounds,* such as tributyltin oxide,

which are no longer used in cooling systems.

The main organic biocides which are currently most widespread on the market are formulated with the following active molecules:

- aldehydes: formaldehyde and glutaraldehyde(toxic)
- the 2.2-dibromo-3-nitrilopropionamide (DBNPA)

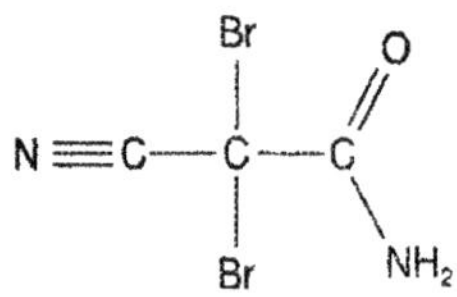

- isothiazolones: in particular, the formula: 5-chloro-2-methyl-4-isothiazolin-3-one/2-methyl-4-isothiazolin-3-one, which is the chemical association that is most widely used in industrial water. They are very effective on a large spectrum of micro-organisms. They are also compatible with chlorinated and brominated treatments.

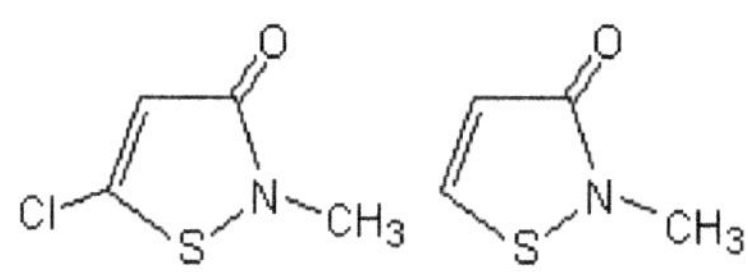

- tetrakis hydroxymethyl phosphonium sulfate (THPS): it has a large spectrum of action. It is very effective for the Legionella and sulphate-reducing bacteria.
- quaternary ammonium salts, characterized by the following structure, where there are four organic substituents on nitrogen,

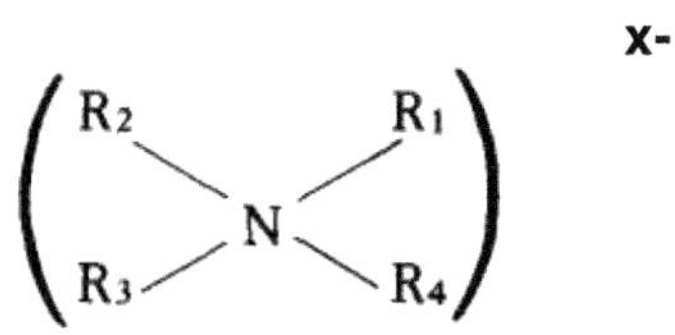

which makes it cationic. The fifth bond is a neutralization electrovalence. In order for these compounds to be bactericidal, they must contain between 8 and 18 carbon atoms. Many quaternary ammonium salts are commercially available. They are very effective for the control of algae development. Certain compounds may produce foam during their injection into the circuit. As they are cationic, they are incompatible with anionic polymers used as dispersing agents. The main commercial products are chlorides of alkyl trimethyl ammonium and alkyl dimethyl benzyl ammonium.

The non-oxidizing biocides are in general less corrosive than the oxidizing biocides. They are used mainly in industrial water systems. They are often used in association with biodispersants or surfactants, which improves their effectiveness. They act on microorganisms by disrupting their metabolism. They are effective over a wide pH range. In order to avoid the adaptation phenomena, we often inject, in the form of periodic addition, two different biocides alternately.

Commercialization of biocide products is subject to the directive 98/8/CE of the European Parliament.

9.5.3 Biodispersants

Biodispersants are organic molecules of the family of wetting agents. They modify the surface tension of the water and favor: the dislocation of biofilm and penetration of biocides (Javel water, bromine, synthetic bactericides) into biological deposits. They are not toxic. They can be used continuously in treatments of cooling systems. They are also used in curative treatment in the cleaning operations.

These agents can be anionic, non-ionic or even cationic. The ionic character must be known in order to avoid problems of compatibility with other products of treatment present in the water.

9.5.4 Modes of Treatment

Bactericidal treatments can be implemented in 2 different ways:

- continuous
- discontinuous

The objective of any bactericidal treatment is to permanently maintain the Legionella concentration lower than 1,000 CFU/l. The treatment program which is defined by choosing an active matter, its concentration to be applied and its mode of injection, must meet this objective.

The active matters can be classified in two categories:

- oxidizing products: Javel water, bromation, chlorine dioxide, etc.;
- organic products: isothiazolones, 2.2-dibromo-3-nitrilopropionamide (DBNPA), tetrahydroxyphosphonium sulfate (THPS), etc.

The oxidizing products are characterized by absence of remanence: rapid disappearance of the active matter. Therefore, only a continuous treatment will allow meeting the objective of permanent action.

The organic products are characterized by:

- a minimal dose in order to obtain the Legionella mortality rate of 100%;
- the half-life time of the active matter in the installation, which means that the acive matter may degrade naturally in the cooling system water. This is an advantage in terms of environment protection, and it is an element to be taken into consideration in terms of effectiveness;
- relatively high cost, in general.

These elements taken into account, the discontinuous treatment with organic bactericides is the most frequently used method of treatment. In case of such treatment, the quantity injected each time should be such that it must ensure a concentration of the active matter higher, at any moment, than the minimal concentration meant to obtain the 100% mortality of bacteria. The dosage must thus take into account the following:

- the (half) holding time of the water in cooling system considered;
- the half-life time of the active matter suggested in the conditions of use.

The professional (manufacturer) must specify it in its offer. The manufacturer must also specify the concentration to be maintained, expressing it in ppm of the acitve matter.

Currently, there exist kits of dosage, which allow dosing most active matters suggested and thus controlling the presence of the active product.

9.6 Other Treatment: UV

Sun has a well-known purification effect thanks to the emission of an invisible light: the ultraviolet ligh.

This natural phenomenon has been used and amplified in industrial equipment, the UV sterilizers, whose principle is to generate ultraviolet rays in a treatment chamber. These rays inactivate the living cells contained in the liquid passing through the equipment. The DNA and RNA of the microorganisms (virus, bacteria, yeasts, fungi, etc.,) are affected by the UV wavelengths (200 to 315 nanometres). This degradation blocks the replication of the genetic material and prevents the microorganism from surviving or multiplying.

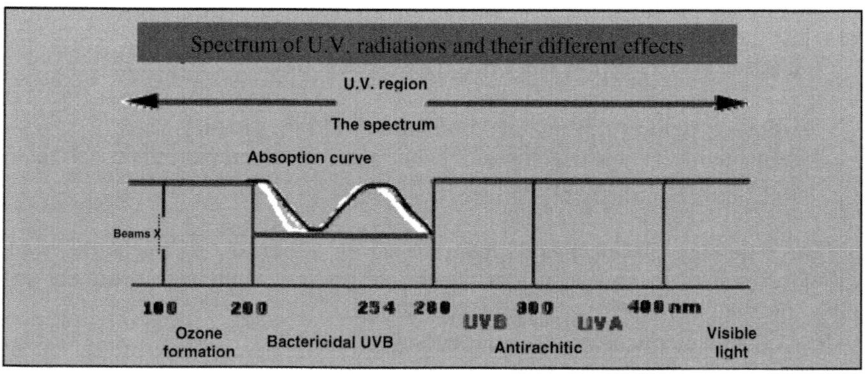

Spectrum of bactericidal activity of UV rays

Currently, this technique is not much used to control microorganism proliferation in industrial cooling water systems.

10

Start-Up of Cooling Installations

10.1 Introduction

The start-up of new recirculating water cooling tower system is an important operation which must be implemented with care.

In fact, the result of a bad procedure of start-up may be a considerable reduction of heat exchanger service life and difficulties in maintaining the heat transfer initially considered.

The problems may appear in only several weeks after the start-up.

However, well programmed start-up will allow an operation without problems and a longer service life of the materials of equipement used.

Thus, the preparation and implementation of the initial start-up of the installation is very important. This preparation is based on knowledge of the system, its design, its structure and of the treatment to be used during the first critical weeks of its functioning.

As we have previously discussed, all cooling systems are based on water circulating through all its elements. Thus, everything that the water touches, removes, transports and deposits can materially affect one or several of the numerous components of the system.

Therefore, an individual component may cause serious damage to other components belonging to the system, due to the transmission capacity of the circulating water. Processing of cooling water in order to prevent corrosion, deposits, algae, etc., utilizes this capacity to transport chemicals to the points that must be protected.

Let us now examine what often occurs during the initial start-up of a system with cooling tower.

After building the cooling tower and connecting the associated piping and conduits, and after installation of the exchangers, etc., the exchangers are water-tested and rinced. The system is then filled with water. The water often remains stagnant for a period of time, awaiting termination of tower construction, and it may not be chemically treated.

The system is then started up, usually under little or no load/pressure, in order to monitor it in terms of leaks, etc.

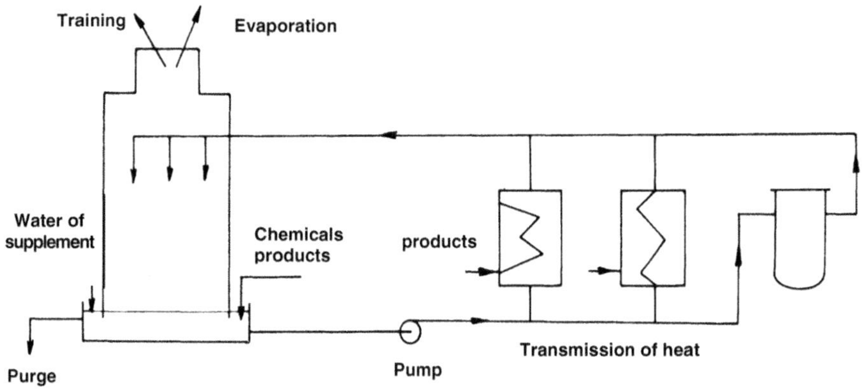

Diagram of recirculating water cooling tower system

The water is recirculated through the cooling tower and the installation. The water contacts the construction materials of the tower. At the contact with concrete, it extracts the soluble substances and distributes them throughout the cooling system. These soluble matters may settle in distribution piping or exchangers, or may react with other subtsances present in the water, then settle in the equipements. This is valid for all substances present in the piping, exchangers or reservoir of the tower. These matters are carried over by the water, and they are either dissolved or transported throughout the cooling system.

Frequently, pieces of wood, paper, plastic and sometimes other objects accidentally left in the system during its construction may be found in the heat exchangers.

There are several important and critical sectors to be considered separately during the system start-up. These are:

- the cooling tower;
- protection of the heat exchangers and other equipements;
- controls of water treatment.

10.2 The Cooling Tower

Cooling towers are important, even essential, parts of the cooling system. When considered separately, the cooling tower receives much less attention than the heat exchangers, to certain extent, because the tower is a component not directly involved with the product manufactured or the industrial process. If the tower becomes scaled or covered with algae, this does not affect the cooling of the manufactured product as rapidly as if it had occurred in the heat exchangers. Similarly, less attention is paid to the construction materials than to the other elements of the system. Wood is no longer used since many years. It has been replaced by steel and concrete. The location of the tower and its environment are also important factors to be considered: the tower functions as an "air scrubber" and would collect dust, spores and other particles, which are found in the circulating water, including bacteria of the Legionella type.

10.3 Protection of Heat Exchangers

In a recirculating water cooling system, it is the equipments to be cooled that are the most sensitive to corrosion and deposits (these components may be tubular exchangers with their water box and tubular plates, plate exchangers, the double-wall reactors or the condensers, to mention only the most common). This is particularly true of the equipments which are often made of carbon steel, but is also applicable to other often used metals and alloys.

Often, these components in a new system (although replacements or cleaned parts of an old system must not be excluded from this situation) do not receive, at the beginning, a sufficient protection. This may cause rapid corrosion, fouling, thus a premature replacement of the components concerned. In order to protect the equipments from corrosion, it is not sufficient to start up with clean surfaces, but it is also necessary to use an inhibitor adapted in order to rapidly and effectively protect the metallic surfaces. A new cooling tower system is often tested with untreated water or with water which received very little treatment. The heat exchangers then remain in contact with the stagnant water for a period of time sufficient for corrosion and corrosion products to appear.

The manufacturing and installation of heat exchangers are a source of accumulation of lubricants, rust, etc., which must be removed before the system start-up. This protection must also be compatible with the corrosion inhibition program which will be applied after the start-up.

Usual start-up of a water cooling system consists in rinsing the exchangers with water. This does not eliminate most foreign matters. After rinsing, an effective corrosion inhibitor is often added to the system water, at a normal maintenance concentration (or several times higher), in order to establish protection. These concentrations may not be sufficient to ensure good protection, even on clean surfaces. An intense localized or pinhole (pitting) attack, on dirty surfaces, may lead to perforations, thus leakages, and a reduced heat transfer, requiring cleaning and repair several weeks after the system start-up. Such effects have been reported in all types of equipments and operation conditions.

In order to illustrate the effects of corrosion, laboratory tests have been carried out at the same time on heat exchanger tubes made of mild steel and on corrosion coupons, using high doses of different corrosion inhibitors considered to be very effective products to protect against corrosion. The results showed that if the same inhibitors were used in parallel at normal maintenance doses and at doses 3 - 6 times higher, on clean carbon steel surfaces, corrosion and tuberculation were controlled effectively only with the high doses, and not with the normal maintenance doses.

The experience also showed that the carbon steel heat exchanger tubes (with water circulating outside the tube bundle) which have been used as is, that is to say, still partially covered with oil, etc., underwent a significant tuberculation and deposit formation, with a localized and pinhole attack underneath, because they had not been properly cleaned before applying the first treatment.

In order to correctly start up a cooling water system with tower and to prevent this type of deposit and attack, a chemical preparation operation is necessary. It is implemented in two stages:

— a complete cleaning;
— the formation of a protective film.

10.3.1 Cleaning

a) *Acid cleaning*
An effective method to remove iron oxide (rust), etc., consists in recirculating, in the equipments, a concentrated solution of passivated acide. In general, a cleaning with a surfactant or an alkaline agent first removes oils, tube drawing compounds, etc., exposing the deposits which will be removed by the acid.

The acid used for cleaning will be chosen depending on the nature of the metals present in the system: hydrochloric acid for ordinary phosphoric steel and stainless steels, the sulphamic acid for aluminium and galvanized iron.

Organic phosphorus acids and esters are also used often, especially when they have wetting properties in relation to the metals.

b) *Non- acid cleaning*
Usual method of non-acid cleaning consists in using a mixture of a sequestering agent and a surfactant, to remove fats and oils, as well as smudges and rust. Phosphate cleaning products act at the same time as degreasing agents and rust remover.

Sequestering agents used are mainly sodium salts of ethylene diamine tetra acetic acid (EDTA) or those of gluconic acid.

c) *Alternating cleanings*
In certain cases, two treatments are alternated: first, cold acid treatment, which is very active, followed by hot alkaline treatment (it is slower). The latter treatment has the additional advantage of perfectly neutralizing any residual acids, which must be totally neutralized in all the parts of the installations. Absence of such a treatment may cause pitting attacks in poorly rinsed regions.

10.3.2 Protection

Protection must assure the most rapidly possible formation of a continuous protective film.

This notion is very important for understanding the use and successful control of corrosion inhibitors commonly used in cooling systems. The inhibitor concentrations normally monitored in cooling water are the maintenance concentrations, which maintain the already formed protective film on metallic surfaces. The establishment of this protective film, however, requires concentrations 3 - 5 times higher, and this film will be maintained for several days.

This is crucial for ensuring a good protection and avoiding attack on metallic surfaces, which can be more serious than in the absence of inhibitor. This method must be applied to new or recently cleaned heat transfer material, or when the protective film has been destroyed for any reason, during operation.

The following doses are usually recommended for the monofunctional inhibitors listed:

Inhibitor	Start-up/maintenance doses
Polyphosphate	40 - 60 ppm/10 –30 ppm
Zinc	10 - 20 ppm/3 - 5 ppm
Polysilicate	40 - 50 ppm/10 –20 ppm
Molybdate	40 - 60 ppm/5 - 20 ppm
Phosphonate	10 – 20 ppm/ 5-10 ppm

These doses are maintained during several days when the installation is started up. However, when one particular component of the system is replaced or cleaned (this is usually the case for the components operating in the most difficult conditions) even higher doses are used on an isolated element, but only for several hours.

The use of phosphate based inhibitors provides much better protection, since no pitting corrosion is produced and an additional elimination of iron oxide is noted. Usually, 50 ppm of phosphate is required for the entire system, whereas high concentrations will assure protection in several hours. These inhibitors are perfectly compatible with continuous inhibition programs.

The use of other types of inhibitors must be examined in terms of protection or attack of various metals that may be encountered in a cooling tower, and in terms of the compatibility with the other inhibitors in the system, so as to assure continuous control and effectiveness. Currently, only copper based inhibitors are used in systems containing primarily copper alloys

10.4 The Control of Water Treatment

There is another important consideration concerning the start-up of new systems. The effort for obtaining an effective protection of the equipment at the beginning of start-up would be totally useless, if no serious controls of the water treatment are carried out. Many plants are often reluctant to initially install control equipment, for cost reasons. Subsequently, a pH drop which will "acid wash" the system (or poorly maintained inhibitor concentrations which may reduce the system protection), will destroy the initial conditioning and cause the onset of corrosion. This of course is not done intentionally, as the attention is usually focused on the product manufacturing. Manpower is often not available, or is available too late, after the damage has already occured. If the controls listed below are carried out at the beginning and assured during the start-up of a new cooling system, they will continue maintaining, during long time, the best possible protection with treatment program adopted:

- precise control of pH;
- maintenance of inhibitor doses at 2 - 3 times their normal concentration levels;
- addition of biocides, even if there is no apparent need;
- continuous supply of chemicals;
- maintenance of cycles of concentration;
- maintenance of good flow rates throughout the installation;
- surveillance of rates of corrosion;
- complete initial conditioning of the system;
- immediate action, if there is an accidental contamination by the process (leakage).

There are many equipments which facilitate start-up, control, as well as monitoring of the system treatment. Automation of the system, continuous monitoring of the parameters and teletransmission of information allow constituting records of information concerned, in order to determine the actions to be taken and alarm setting for a more rapid intervention.

10.5 Conclusion

The correct start-up of new cooling systems requires the knowledge of potential problems, the capacity to localize these problems, as well as the capacity to take appropriate decision in order to prevent or minimize these problems. It is also necessary to examine all the elements of the cooling system concerned, individually and together, particularly the equipement of the cooling tower, as well as that of the heat exchangers. It must be remembered that it is the water that unites the system as a whole, and it causes and transports the problems, as well as their solutions.

APPENDIX

1

Water Analysis

The most important criterion of choice, when searching for an appropriate treatment, is the composition of the water available. Moreover, the analysis of treated water allows monitoring good functioning of the treatment applied.

The purpose of the cooling water monitoring is to avoid the accidents described in the second part of the book, by maintaining the concentrations of dissolved salts, phosphates, silica, etc. within the limits required for good protection.

The type and frequency of the analyses to be carried out vary depending on operating conditions and the observed incidents of functioning. In particular, in the case of large cooling installations, frequent analyses are not necessary, given the mass inertia related to the considerable volume of basins, in which concentrations change very slowly. In general, daily measurement of the calcium hardness, of the concentrations of chlorides, dispersing agents and corrosion inhibitors, as well as that of pH are sufficient. Bacteria counts must be performed every week.

The choice concerning analyses and thier frequency should be made in collaboration with the water specialist, which will take into account the specificity of each system.

Methods of the Legionella concentration analysis must be chosen in accordance with the relevant legislation in force, concerning the respect of standards imposed. The frequency of the analyses will depend on the level of Legionella concentration measured.

The analyses should be carried out according to the method NFT 90-431.

Within the framework of the Legionella risk management, the relevant analyses can be performed by using the PCR (polymerase chain reaction) method.

Many analyses can be performed on site by relatively unqualified personnel using simple methods. There exist special analysis kits for water treatments, including the material and reagents required for the main analyses (titrimetry and colorimetry).

It is nevertheless advisable to carry out periodic verification of the validity of results obtained using routine on site determinations, by qualified personnel in the laboratory. Moreover, some analyses are possible only in laboratory conditions.

A certain number of factors affect the precision and the validity of the results:

— *Reagents:* they must always be prepared in a well equipped laboratory by qualified personnel. Their reactivity must always be verified and they must not be used after their expiration date.
— *Sampling procedure:* it is important that the sample be representative of the water to be analyzed. Therefore, sampling points must be chosen. The recipient should be rinsed several times with the water to be analyzed before filling it with the sample. When a sample is taken from a valve or a faucet, the water should run several minutes before collecting the sample. Only clear samples are analyzed. It is advised to identify by a panel the place of sample chosen.
— *Temperature*: certain measurements, such as those of pH and conductivity, are temperature sensitive. Ambient temperature (20-25°C) is suitable for most analyses.

The analysis kits are commercially available and the methods used are interesting for their simplicity, rapidity and facility of implementation, while offering good precision and reproducible results. They are sufficient for the most common analyses necessary for the installation and control of water treatment in a cooling circuit. In the case of less frequent analyses or those required by unusual circumstances, we will refer to the complete works, such as “L'analyse de l'eau” (“Water Analysis”) by J. Rodier (Ed. Dunod) and “Chemical Analysis of Industrial Water” by J.W. Mc Coy (Chemical Publishing Company N.Y.).

Devices of measurements or analyses allow realizing continuous analyses, or in any case, with a frequency which can be fixed beforehand. This concerns the measurements of, for example, the pH, temperature, conductivity, corrosion, free chlorine, etc.

Recording of the measurements allows creating large data banks helping to assure an optimization of treatments and reveal possible disfunctions.

2

Water Treatment Glossary

Acid: Substance which is dissociated, more or less completely, in aqueous solution, producing hydrogen ions H+. An acid is all the stronger since it gives, in the water, more ions H^+. Among the acids used in water treatment, we can cite: sulfuric acid (H_2SO_4); hydrochloric acid (HCI).

Acidity: Concentration of hydrogen ions H^+ in a solution. A solution is all the more acid since its pH is low (below 7).

Softening: Treatment process meant to eliminate water hardness (due to the presence of alkaline-earth salts: carbonates, sulphates, and calcium and magnesium chlorides). Softened water is not scale-forming. The softening is carried out by passing water on a cation exchange resin (permutation of calcium ions with sodium ions) regenerated with sodium chloride.

Adsorption: Physical process of adherence of the substances dissolved or dispersed to the surface of a solid body (see activated carbon).

Aerobic: Concerns bacteria which develop in the presence of oxygen of the air, free or dissolved.

Aggressiveness (of water): Aptitude of water to dissolve scale, related to the presence of excess carbon dioxide in the water (aggressive carbon dioxide). The scale dissolving continues until the calcium carbonate stability is reached. Aggressive water can not form calcareous deposit. The aggressiveness of any water decreases as its temperature increases. The aggressiveness of water is measured by marble test. (Not to confuse with corrosivity).

Alkalinity: Concentration of hydroxyl ions OH^- in a solution. A solution is all the more alkaline since its pH is high (higher than 7).

Alkaline-earth: Group of bivalent metals including, among others, calcium, magnesium and barium. Their salts dissolved in any water constitute its hardness.

Alkalines: Group of monovalent metals including, among others, sodium and potassium. They make part of the composition of so-called strong bases (or alkalis). (caustic soda, potash).

Anaerobic: Concerns bacteria which develop in the absence of oxygen of the air, free or dissolved.

Anion: Negatively charged ion (examples: Cl^-, SO_4^{--}, HCO_3^-, OH^-).

Anode: Positive electrode of an electric current generator. The anions go towards the anode in an electrolyte system.

Make-up water: Water introduced into a cooling circuit in order to compensate for the losses due to the evaporation, possible leakages and blow-downs.

Base: Substance which is dissociated, more or less completely, in aqueous solution, producing hydroxyl ions OH^-. A base is all the stronger since it gives more ions OH^-, in the water.

Bases react with acids, producing a salt and water.

Among the bases used in water treatment, we can cite: lime $Ca(OH)_2$, caustic soda NaOH and potash KOH.

Ionic balance of water: Balance in two columns (cations and anions) of the results of water analysis, while the concentrations are expressed in milli-equivalents per litre or in French degrees. Here, the total amounts must balance each other out, since in a saline aqueous solution, the sum of cations is equal to that of anions.

Calcium bicarbonate or calcium hydrogen carbonate: Salt present in water of calcareous origin and which is one of the causes of its hardness. Given the formula $Ca(HCO_3)_2$, it is instable in aqueous solution and tends to decompose into $CaCO_3$ and H_2CO_3 (carbonic acid). Heat accelerates this transformation which leads to the scale deposit (scale) on the surfaces of heat exchangers, as well as in the piping.

Biofilm: A biofilm is a bacterial community adhering to a surface. Excessive production of biomasses fixed on the metallic surfaces in contact with water, associated or not with mineral deposits, is the cause of problems of organic fouling and corrosion of bacterial origin. The biofilms are also known for being cause of bacterial contaminations of the circulating water by liberating the adhesive bacteria. This is the case of the Legionella.

Bromation: This is a technique of disinfection, which is used more and more in industrial cooling systems, as it presents many advantages compared to the chlorination (Javel water):

Superior performances when used for alkaline waters;
Less corrosion, because the dosage of oxidizing agent is lower.

Cathode: Negative electrode of an electric current generator. The cations go towards the cathode in an electrolyte system.

Cation: Positively charged ion (examples: Ca^{2+}, Na^+, H^+).

Chloramines: Compounds of substitution formed by reaction of chlorine Cl_2 and ammonia NH_3 (example: monochloramine NH_2Cl). In water chlorinated at a dose inferior to the critical point, the residual chlorine is found entirely in the form of chloramines (or bound chlorine). Chloramines are sometimes used for disinfection, when slow and prolonged action is desired.

Chlorination: usual technique of disinfection used for controlling the development and proliferations of microorganisms, which consists in injecting into the circuits the products liberating active chlorine such as Javel water, chloroisocyanuric acids and chlorine dioxide. Avantage: its effectiveness in relation to its cost. Disadvantages: corrosive on steel and not very effective to the alkaline pH. In the latter case, bromation is preferred compared to chlorination.

Clarification of water: All of the treatments meant to remove suspended solids, turbidity and colour of any water. Clarification of water is usually carried out according to one of the following plans:

- direct filtration;
- partial coagulation followed by in-line flocculation;
- coagulation, flocculation, settling, filtration.

Coagulant: Chemical product meant to favor the coagulation of suspended colloidal matters in water. The most frequently used ones are aluminium salts such as ferric chloride $FeCl_3$, sulphate and polychlorides. Organic coagulants, in particular polyamines and polydadmac, have appeared on the market more recently.

Coagulation: Reduction or removal, under the action of "coagulants" such as aluminium sulphate or certain cationic polyelectrolytes, of negative electric charges carried by colloidal particles suspended in water, making their agglomeration, flocculation and precipitation possible.

Colloid: Particle of very small size (1 to 100 μm or 0.001 to 0.1 mm) suspended in a liquid, also called micelle. The coagulation-flocculation has the effect of separating the colloids from the liquid in which they are suspended.

Concentration of dissolved salts: Progressive increase of the concentration of dissolved elements in recirculating water because of the partial evaporation of the circulating water in a cooling tower.

Conditioning of water: Chemical treatment applied to the waters of cooling systems. The conditioning of water consists in introducing into the circulating water the specific chemicals allowing prevent the formation of scale deposits and control corrosion and development of microorganisms.

Conductivity: Aptitude of any water to allow the passage of electric current. Conductivity is the reverse of resistivity. It increases depending on the concentration of dissolved salts. It also depends on temperature. It is usually expressed in microsiemens per cm ($\mu S.cm^{-1}$).

Corrosion: Attack on a metal surface due to the electrochemical action in an aerated medium (corrosion by oxygen) or non-aerated one (corrosion by hydrogen). It may also be of bacterial origin. The deterioration of the metal surface by abrasion or erosion is due to a mechanical or physical action, too high a speed of passage, presence of suspended solids.

Corrosivity (of water): Aptitude of any water to dissolve metals, related to its physical-chemical composition (pH, resistivity, concentrations of oxygen, chlorides and sulphates). Corrosivity of water increases depending on temperature increase.

Settling: Physical process of separation of suspended solids in a liquid, using the gravity action in order to collect them totally or partially and make them gather at the bottom of a reservoir (clarifier). In the case of water having high concentration of suspended solids, the settling is frequently used prior to the filtration.

Carbonate removal: Process of water purification in order to remove bicarbonates that it contains in dissolved state:

Through precipitation by lime, if it is a matter of alkaline-earth bicarbonates, with formation of insoluble carbonates; through the ion exchange, with the release of carbon dioxide, chlorides and sulphates remaining in solution.

Simply decarbonated water is softened only partially, since it may still contain calcium and magnesium salts other than carbonates.

Chlorometric degree: Unit of active chlorine concentration of Javel water, commercial name of sodium hypochlorite. One chlorometric degree corresponds to 3.17 gramms of active chlorine per litre.

French degree: Unit of concentration of chemical substances in aqueous solution. Used in the water treatment practice, the French degree corresponds to the concentration of a solution N/5000.

One French degree (1°) is equivalent to 0.2 milliequivalent per litre.

This unit is used, in particular, to express:

- calcium hardness, magnesium hardness and total hardness (TH);
- phenolphthalein alkalinity (PA);
- methyl orange alkalinity (MA).

Correspondence with other frequently used degrees:

$$1°F = 10 \text{ ppm } CaCO_3 = 0.56° \text{ German} = 0.7° \text{ English.}$$

Titration for hardness or TH: Unit of hardness (concentration of alkaline-earth ions) of water, expressed in French degrees.

Demineralization: Process of water purification in order to totally remove the salts dissolved in it. The demineralization is carried out:

- through passage on ion exchange resins;
- through reverse osmosis and through distillation.

Disinfection: Operation allowing reduce by 10^5 the microbiological population of water (see also sterilization).

Hardness (of water): Concentration of calcium and magnesium or total hardness.

The hardness or titration for hardness TH (or total hardness) indicates on the whole the concentration of calcium ions Ca^{2+} and magnesium ions Mg^{2+}.

The calcium hardness expresses the concentration of calcium ions Ca^{2+}.

The magnesium hardness expresses the concentration of magnesium ions Mg^{2+}.

The hardnesses are expressed in French degrees.

Carbonate hardness indicates the concentration of calcium and magnesium ions susceptible to precipitate in the form of carbonates. It is equal to MA, if the TH is superior to MA, or equal to TH, if the TH is inferior to MA.

Temporary hardness indicates the concentration of calcium and magnesium, which precipitate after prolonged boiling of water.

Permanent hardness indicates the concentration of calcium and magnesium, which remain in solution after prolonged boiling.

Javel water: Aqueous solution of sodium hypochlorite (NaClO) used, in particular, for disinfection of water. This is a powerful oxidizing agent. The concentration of Javel water is expressed in chlorometric degrees: one chlorometric degree corresponds to 3.17 gramms of active chlorine per litre. The technical commercial solution titrates 47 to 50°chlorometrics corresponding to about 150g/l in equivalent of active chlorine.

Anion exchanger: Ion exchange resin susceptible to fix mineral or organic anions and exchange them, either between them, or with hydroxyl ion OH^-.

We distinguish:

- weakly or moderately basic exchangers, containing a mixture of amines;
- strongly basic exchangers containing radicals of quaternary ammonium.

Cation exchanger: Ion exchange resin susceptible to fix mineral or organic cations and exchange them, either between them, or with hydrogen ion H^+. We distinguish:

- strongly acid exchangers characterized by the presence of sulfonic radicals: $-SO_3H$;
- weakly acid exchangers characterized by carboxylic radicals: -COOH.

Eluate (in ion exchange): Liquor evacuated by an ion exchange column during an operation of regeneration.

Scale formation or scaling: Formation, on the surface of heat exchangers and other equipments, as well as in the piping, of hard and adhesive scale deposit constituted mainly by calcium salts. It increases with the temperature increase.

Calcium carbonate stability: State of water whose concentrations of calcium, free CO_2, bicarbonates and carbonates are such that, when entering in contact with scale, its characteristics do not change: there are many graphic methods to determine the pH of equilibrium or pH of saturation (pHs), which allows then calculate the scale-forming or corrosive power of any water whose composition is known at a given temperature.

Slow filtration: Filtration without prior coagulation or settling on granular material (most often sand or anthracite) of water which is moderately rich or very rich in suspended solids. Low speed of filtration, 2 to 5 $m^3/j.m^2$, allows quite reduced head loss. In order to obtain good results, several stages of filtration are required.

Rapid filtration: Filtration with or without coagulation (direct) on granular materials, of water having low or very low concentration of suspended solids, at a speed located between 4 and 50 $m^3/h.m^2$.

Flocculant: Mineral or organic product meant to favor the flocculation of the matters coagulated beforehand, present in water to be treated. We distinguish: mineral flocculants (activated silica, etc.), organic flocculants of natural origin (alginates) and synthetic flocculants (polyelectrolytes of the type polyacrylamide).

Flocculation: Agglomeration and precipitation of colloidal particles suspended in water and coagulated beforehand. This agglomeration is facilitated by the addition of flocculants into the water. The use of new synthetic polymers has allowed considerably improving the flocculation performances.

Fouling index: It represents the fouling potential of water. This measurement is carried out through determination of the fouling degree of a filter.

Halogens (from Greek: "salt generator")**:** Elements of the VII group of the periodic table of elements: fluorine, chlorine, bromine, iodine.

Hypochlorite: Chemical compound containing the radical ClO^-. Hypochlorites are used, as chlorine, for disinfection of cooling waters and for the control of algae and bacteria developement in industrial water circuits and domestic hot waters. Following products are commercially available:

- liquid: sodium hypochlorite or Javel water;
- solid (powder): calcium hypochlorite.

Both of them are used after dilution or dissolving in water.

Ryznar index: Index allowing quantify the scale-forming or aggressive aspect (in relation to marble), of water without treatment.

In a cooling circuit, it is located between 2 values which depend on minimum and maximum temperatures of skin, to which the circulating water is subjected.

Its knowledge is involved in the choice and the dosage of the scale-preventive and anticorrosion formulation applied.

The Ryznar index is equal to 2 pHs (pH of saturation)- pH.

The calculation of the pH of saturation is issued from a polynomial formula, and the logarithmic part most influences its value. The formula given below provides a good approximation of it.

pH of saturation = $10.287+(1.448\cdot10^{-4}$ x salinity)-(0.0197xT°)- log10(THCa x MA),

where: salinity is expressed in mg/l,
temperature expressed in ° Celcius,
calcium hardness THCa and MA expressed in French degree.

Corrosion inhibitors: Chemicals which, added to water, prevent its corrosive action on metals through formation of a protective layer on the metal.

Ion: Electrically charged atom or group of atoms, presenting an excess (anion) or a deficit (cation) of electrons, whose migration causes the passage of electricity through an electrolyte.

Positive ions or cations include: metals and hydrogen (example: Ca^{2+}, Na^{+}, H^{+}).

Negative ions or anions include: non-metals and groups such as SO_4^{2-}, HCO_3^{-}, etc., and hydroxyl ion OH^{-}.

Floating bed: Process of setting up the ion exchangers where the production cycle is realized upwards and the regeneration - downwards.

Suspended solids: Solid particles of very small size, present in water. We distinguish: settleable solids, which are separated naturally due to gravity, when water is at rest; and colloidal matters which are too fine for settling and require addition of coagulant for their removal.

Milliequivalent per litre (meq/l): Unit of concentration of dissolved matters in an aqueous solution. 1 meq/l corresponds to the concentration of a standard electrolyte solution considered to be diluted thousand times. This is the concentration of a solution N/1,000. The meq/l is equivalent to 5 French degrees.

1 meq/l = 5°F

Molarity (of an aqueous solution): Number of moles of solute contained in one litre of water.

Nanofiltration: Tangential or dead-end filtration on synthetic membranes (mineral or organic) leading to the total removal of suspended solids and partial removal of the largest dissolved elements.

Neutralization: Operation which consists in bringing the pH of a solution to a value close to 7: by acidification, if its initial pH is higher than 7, and by alkalization, if the initial pH is lower than 7.

Normality: A normal aqueous solution is one that contains one gram equivalent of the solute considered per litre of solution.

Osmosis: Natural diffusion of a solvant through a semi-permeable membrane, from diluted solution towards the concentrated solution. The difference of concentration generates a pressure called "osmotic", and it has the effect of equalizing the concentrations on the both sides of the membrane, thus diluting the most concentrated solution. The osmosis plays a major role in the circulation of water in living organisms.

Reverse osmosis: Process of membrane separation meant to extract a solvant from a solution (example: pure water from salt water). It consists in (by reversal of the natural osmosis process) applying to a solution in contact with a semi-permeable membrane, a pressure superior to the osmotic pressure, and collecting the solvant from the other side of the wall. This process uses pressures of 3 to 100 bars and allows removing the particles of the sizes of 0.2 to 10 nm (almost all the ions and solutes).

pH (potential hydrogen): the pH is defined as being the decimal cologarithm of the concentration of ions H^+ of a solution expressed in mole/l.

The pH varies between 0 and 14; whereas 7 is the pH corresponding to the neutrality. Water is all the more acid since its pH (lower than 7) is closer to 0, and all the more alkaline since its pH (higher than 7) is closer to 14.

The pH varies depending on temperature and is measured through a pH meter.

Cathodic protection: Method which consists in placing, in the circuits to protect, the electrodes of a metal whose potential is lower than that of the metal to protect. They play the role of soluble anodes in relation to the metal which becomes cathode. This type of protection is particularly used for underground piping, the metals in contact with sea water. It is difficult to implement in an industrial system.

Blowdown: Discarding of a part of the total water volume contained in a water system, in order to limit the increase in concentration of dissolved salts of the circulating water due to the partial evaporation in the form of pure water. The losses of water due to the blowdown and the evaporation are compensated by the make-up water.

Regeneration (in ion exchange): Operation carried out on a saturated ion exchange resin, so as to bring it to its initial state. Regeneration consists in making percolate on the resin the appropriate solution (acid, basic or saline) of high purity. The regeneration is always followed by a slow rinsing, then by a rapid one.

Co-current regeneration: Process of regeneration of an ion exchanger, which consists in making circulate the regenerant solution from top to bottom of the resin bed.

Counter-current regeneration: Process of regeneration of ion exchangers, which consists in making circulate the regenerant solution in the direction opposite (from bottom to top of the resin bed) to that of the water to be treated. The efficiency is thus better than that in the case of co-current regeneration.

Ion exchange resin: insoluble resin presented in the form of granules or spheres, usually organic and synthetic, containing in its structure the acidic radicals (cation exchangers) or basic radicals (anion exchangers), and susceptible to change over, without being affected itself, the positive or negative ions fixed on these radicals, against the ions of the same sign in solution in the liquid with which it enters in contact. These resins are placed in the ion exchange columns, where the composition of the water to be treated is modified, through percolation. These resins are regenerated.

Applications: softening, carbonate removal, demineralization.

Resistivity: Resistivity expresses the resistance of water to the electricity passage. It is the reverse of the conductivity. It is measured in ohms.cm (Ω.cm). The relation between the two measures is the following:

$$\text{Resistivity in ohms}\cdot\text{cm} = 10^6/\text{conductivity in } \mu\text{S}\cdot\text{cm}^{-1}$$

SAS: Total concentration (expressed in French degrees or in meq/l) of strong acid salts (chlorides, sulphates, nitrates) present in a solution.

Salt: Substance resulting from the action of an acid on a base. Among the salts used in water treatment, we can cite:

- sodium chloride $NaCl$;
- sodium silicate Na_2SiO_3;
- ferric chloride $FeCl_3$;
- aluminium sulphate $Al_2(SO_4)_3$, 18 H_2O.

The salt used for the regeneration of water softeners is constituted by sodium chloride of very high purity.

PA (phenolphthalein alkalinity): Measure of the concentration, in water, of hydroxides and the half of its concentration of alkaline and alkaline-earth carbonates, determined by addition of the quantity of sulfuric acid required for the colour change of phenolphthalein from red into colourless, at pH 8.3.

If the pH is lower than 8.3, the PA is nil and the water practically contains only bicarbonates.

The PA is usually expressed in ppm of $CaCO_3$.

Its French equivalent (TA) is expressed in French degrees (°F). The German equivalent is "pWert", expressed in German degrees (°dH).

MA (methylorange alkalinity): Concentration of alkalis (hydroxides), carbonates and bicarbonates (alkaline and alkaline-earth) in water, determined by addition of the quantity of sulfuric acid required for the colour change of methylorange (or helianthine) from yellow into orangey, at pH 4.3.

The MA is usually expressed in ppm of $CaCO_3$.

Its French equivalent (TAC) is expressed in French degrees (°F). The German equivalent is "mWert" and expressed in German degrees (°dH).

Tartre: Deposit usually hard and adhesive, constituted mainly by calcium salts (carbonates, sulphates, silicates).

Content of dissolved salts: Concentration of elements dissolved in water. The concentration is expressed most often in units of weight per unit of volume (mg/l or g/m³ or ppm) or in units of electric charges of ions per unit of volume (meq/l or °f).

TH (hardness or titration for hardness): See Hardness of water.

Turbidity: Turbidity of water is the reverse of its transparency. It depends mainly on the presence of suspended solids in the water. It is measured by comparison with a sample of distilled water, to which we add drop by drop a solution of mastic, either by measuring the limit of visibility of a submerged object or of an image. It is expressed in drops of mastic, or in Jackson units or NTU.

Ultrafiltration: Process of filtration on synthetic membranes (mineral or organic), situated between the microfiltration and the nanofiltration.

Bibliography

J. CHANTEREAU : Corrosion bactérienne et bactéries de la corrosion.

J.I. BREGMAN : Corrosion inhibition

J. RODIER : L'analyse de l'eau.

AMERICAN PUBLIC HEALTH ASSOCIATION : Standard methods for the examination of water and waste water.

APTEN : Compte rendus des journées information eau 1976-2004

Sté DEGREMONT : Mémento technique de l'eau..

AMERICAN WATER WORKS ASSOCIATION : Water quality and treatment .

Guide de l'eau, édition 2003/2004.

Dr. F. SQUINAZI : Biofilms et matériaux

HACH : Analyse spectrophotométrique des eaux.

J.-F. BOSICH : Prévention de la corrosion.

J.-W. Me COY : The chemical treatment of cooling water.

G. CHARLOT : Les méthodes de la chimie analytique.

J.-L. ROUBATY : Conférence Légionelles (IRH)

Y.H. KIM : Coagulants and Flocculants

L. DUVIVIER: Traitement des eaux dans les centrales thermiques

REPUBLIQUE FRANCAISE : Legionella et tours aéro-réfrigérantes : guide des bonnes pratiques

GPSR Compliance

The European Union's (EU) General Product Safety Regulation (GPSR) is a set of rules that requires consumer products to be safe and our obligations to ensure this.

If you have any concerns about our products, you can contact us on ProductSafety@springernature.com

In case Publisher is established outside the EU, the EU authorized representative is:

Springer Nature Customer Service Center GmbH
Europaplatz 3
69115 Heidelberg, Germany

Batch number: 10682215

Printed by Printforce, the Netherlands